WIRELESS SYSTEMS AND MOBILE COMPUTING

WIRELESS SYSTEMS AND MOBILE COMPUTING

MOHSEN GUIZANI
(EDITOR)

Nova Science Publishers, Inc.
Huntington, New York

Senior Editors: Susan Boriotti and Donna Dennis
Office Manager: Annette Hellinger
Graphics: Wanda Serrano and Dorothy Marczak
Coordinating Editor: Tatiana Shohov
Book Production: Matthew Kozlowski, Jonathan Rose, Jennifer Vogt
Circulation: Cathy DeGregory and Ave Maria Gonzalez
Communications and Acquisition: Serge P. Shohov

Library of Congress Cataloging-in-Publication Data
Available Upon Request

ISBN 1-59033-121-4

227 Main Street, Suite 100
Huntington, New York 11743
Tele. 631-424-6682 Fax 631-425-5933
e-mail: Novascience@earthlink.net
Web Site: http://www.nexusworld.com/nova

The color presentations are on the accompanying CD or diskette.

Printed in the United States of America

CONTENTS

PREFACE

Due to the rapidly expanding technology of wireless and mobile communications, mobile users with portable computers should be capable of accessing information anywhere and at anytime. This new communication style will significantly affect the distributed computing environments.

Mobile and wireless computing is permeating the globe, affecting the way we live and conduct business using portable computers such as laptops, hand-held PC's, Palm computers, and even wearable computers. In becoming mobile, users innocently expect and demand the same computing luxury they used to (and still) enjoy in the fixed computing environment. Unfortunately, network resources (e.g. printers, fax machines, and file systems) and application software do not follow the mobile users when they leave their offices or homes, or when they relocate to another temporary office or home. Supporting true mobility of users and business task forces will therefore require changing the way application software is packaged and deployed. It will also require changing the mechanisms by which system resource are configured and managed. The lesson quickly learned by the mobile computing research community is that mobility and dynamism are synonymous. Service discovery protocols (SDP) are re-shaping the way software and network resources are configured, deployed, and advertised, all in favor of the mobile user. Furthermore, so called *wireless local connectivity* technologies (e.g. Infrared and BlueTooth) are fueling the importance of SDP by enabling the mobile user to discover and use proximity services seamlessly and effortlessly.

In this book, we focused on all aspects of wireless systems that support mobile computing. In particular, all aspects of distributed computing with mobile communication capability. We have received a large number of manuscripts and chose the best 6 of them. These papers span a range wide enough to benefit our readers experts or getting in wireless systems and mobile computing.

The first paper, by C. Lee and S. Helal, introduces the subject of Protocols for Service Discovery in Dynamic and Mobile Networks. The authors discuss the emerging technology of *service discovery* in the context of mobile and wireless computing, with emphasis on Jini, UpnP, and Salutation protocols.

The second manuscript, by Y. B. Lin and J. K. Wey investigate the problem of reducing PCS signaling traffic through delay techniques. This paper describes two delay registration approaches, MS-controlled delay registration (MS-DR) and VLR-controlled delay registration (VLR-DR), to reduce the registration traffic. The authors' results indicate that following appropriate guidelines, these delay registration approaches effectively reduce the number of messages at the cost of losing very few number of call deliveries to a mobile station. Both MS-DR and VLR-DR can be easily implemented without modifying the existing roaming management protocols such as IS-41 and GSM MAP.

Then, L. S. Li and M. F. Chang describe Mobile Two-Tier Gateways (MTTGs) for PCS Systems (High-tier PCS and Low-tier PCS). The authors propose the two types in order to enable low-tier mobile users to use high-tier service so that low-tier users moving at high speed can also be served. The MTTGs can be installed on public transportation vehicles where low-tier service is unavailable due to high moving speed. The roaming management protocols for MTTGs and low-tier mobile users have been developed. The low-tier HLR database has been modified to support efficient registration and location tracking. They evaluate the performance of MTTGs using computer simulations. The results indicate that low-tier mobile users using MTTGs experience higher blocking probability and forced termination probability than high-tier mobile users. The results also show that the two-tier interconnected system using MTTG extends the service area for low-tier users at the cost of degrading the quality of service for high-tier users.

Next, M. Guizani and M. Farahani investigate the mathematical modeling and evaluation techniques to predict the performance of wireless/mobile systems. These techniques are indispensable tool for assessment of the performance of a system, evaluation of design alternatives, resource dimensioning, and system configuration. The authors study the Markov Modulated Poisson Process (MMPP) and use it to model the hand-off calls in cellular systems. An analytical model of a cellular system is solved and the performance metrics are calculated. The results obtained are compared to those obtained using the Poisson model.

Furthermore, S. T. Sheu And M. H. Chen propose a fuzzy bandwidth allocation controller (FBAC) to support real-time traffic over wireless networks. The restricted time-bounded service (RTBS), such as voice and video service is sensitive to packet delay and the loose time-bounded service (LTBS), such as data transfer cares about the accuracy and can be serviced before a pre-assign due time. Based on the FBAC, five different request assignment strategies are introduced. Simulation results demonstrate that the fuzzy controller with some proposed strategies obtain high network performance, high fairness as well as an acceptable blocking probability.

Finally, M. Guizani and W. Alanqar develop and propose a new enhanced Connection Admission Control (CAC) based on learning fuzzy logic in order to monitor and dynamically adjust the reserved bandwidth in adjacent cells where the connection exists. Two types of traffic classes were used to study the learning fuzzy CAC performance, class I is quality of service (QoS) dependent, Class II is QoS independent. It was noticed through the simulation results that learning fuzzy CAC provides less total number of blocked new connections and less total number of dropped mobile connections. This was achieved through the dynamic allocation of reserved bandwidth to cope with the current network conditions.

We hope that this effort can give our readers a good idea of few of the recent active research areas in Wireless systems and Mobile Computing.

Chapter 1

PROTOCOLS FOR SERVICE DISCOVERY IN DYNAMIC AND MOBILE NETWORKS

Choonhwa Lee and Sumi Helal
Computer and Information Science and Engineering Dept.
University of Florida
Gainesville, FL 32611-6120
UNITED OF STATES
{chl,helal}@cise.ufl.edu http://www.cise.ufl.edu/~{chl,helal}

Abstract: - Mobile and wireless computing is permeating the globe, affecting the way we live and conduct business using portable computers such as laptops, hand-held PC's, Palm computers, and even wearable computers. In becoming mobile, users innocently expect and demand the same computing luxury they used to (and still) enjoy in the fixed computing environment. Unfortunately, network resources (e.g. printers, fax machines, and file systems) and application software do not follow the mobile users when they leave their offices or homes, or when they relocate to another temporary office or home. Supporting true mobility of users and business task forces will therefore require changing the way application software is packaged and deployed. It will also require changing the mechanisms by which system resource are configured and managed. The lesson quickly learned by the mobile computing research community is that mobility and dynamism are synonymous. Service discovery protocols (SDP) are re-shaping the way software and network resources are configured, deployed, and advertised, all in favor of the mobile user. Furthermore, so called *wireless local connectivity* technologies (e.g. Infrared and BlueTooth) are fueling the importance of SDP by enabling the mobile user to discover and use proximity services seamlessly and effortlessly. In this paper we discuss the emerging technology of *service discovery* in the context of mobile and wireless computing, with emphasis on Jini, UpnP, and Salutation protocols.

Key-words: - Service discovery, service discovery protocols, lookup services, mobile computing, Jini, UpnP, Salutation, pervasive computing.

1 INTRODUCTION

With the emergence of wireless ad-hoc networks, specialized information appliances are taking over the technology landscape. These information appliances have been born to aim at supporting mobility, in essence, and hence cooperation among them, since cooperation is an indispensable feature that complements some missing parts in mobile device, compared to conventional, fully-powered computing devices. For this cooperation, several service discovery protocols (SDPs) have been proposed as the part of coordination architectures that ensure device interaction with the ultimate aim of simple, seamless and scaleable device inter-operability. Among emerging SDPs, Jini, Universal Plug and Play, Salutation, and SLP are conspicuous. Before discussing them, let us take a look at how information appliances powered by SDPs can be used in the real life.

Scenario 1: Imagine finding yourself in a taxi cab without your wallet. Fortunately, you have a Jini technology-enabled cellular screen phone, and your cellular provider uses Jini technology to deliver network-based services tailored to your community. On your phone screen, you see a service for the City Cab Company, so you download the electronic payment application to authorize payment of your cab fare. The cab company's payment system instantly recognizes the transaction and sends a receipt to the printer in the taxi. You take the receipt and you're on your way.

Scenario 2: Consider an insurance salesman who visits a client's office. He wants to brief new products and their options to the client which are stored in his Windows CE Handheld PC. Since his handheld PC has wireless network and supports UPnP, it automatically discovers and uses an Ethernet-connected printer there without any network configuration and setup. He can print whatever in his H/PC or from computers in his main office and promote the new products.

Scenario 3: Consider an intelligent, on-line overhead projector with a library client. After identification to the system, the user may select a set of electronically stored charts or other document(s) for viewing. Rather than bringing foils to a meeting, the user accesses them through the LAN server in the library.

Scenario 1 is a Jini demo scenario told by Sun and scenario 2 is a UPnP scenario by Microsoft. The last one is Salutation scenario. At a glance, they seem to talk about the same stories: mobile devices, zero-configuration and impromptu community enabled by SDPs, and cooperation of the proximity network. Without trademarks such as Jini and UPnP, we could hardly know which scenario is told by whom.

Even though they play in the same arena, these three SDPs have different origins, underlying technologies, flavors, and audiences. Since they see the problem at different angles and take different approaches to it, they have pros and cons, especially compared to others. This survey is focused on not their overall architectures but SDPs. Before the comparison of the big three, let us take a brief look at the basic mechanisms, goals, and environments of them.

2 JINI CONNECTION TECHNOLOGY

The purpose of the Jini architecture is to federate groups of devices and software components into a single, dynamic distributed system [2]. Jini systems provide mechanisms for service construction, lookup, communication, and use in a distributed system. Examples of services include devices such as printers, displays, or disks; software such as applications or utilities; information such as database and files; and users of the system.

The heart of the Jini system is a trio of protocols called *discovery*, *join*, and *lookup* [2]. A pair of these protocols – discovery/join – occurs when a device is plugged in. Discovery occurs when a service is looking for a lookup service with which to register. Join occurs when a service has located a lookup service and wishes to join it. Lookup occurs when a client or user needs to locate and invoke a service described by its interface type (written in the Java programming language) and possibly, other attributes. The following steps show what interactions are needed among a client, a service provider, and a lookup service for a service to be used by the client in a Jini community [2] [1].

1. Service provider locates a lookup service by multicasting a request on the local network or a remote lookup service known to it in priori.
2. A service provider registers a service object and its service attributes with the lookup service. This service object contains Java programming language interface for the service, including the methods that users and applications will invoke to execute the service, along with any other descriptive attributes.
3. A client requests a service by Java type and, perhaps, other service attributes. A copy of the service object is moved to the client and used by the client to talk to the service.
4. Then, client interacts directly with the service provider via the service object.

Jini connection technology consists of an infrastructure and a programming model which address the fundamental issues of how devices connect with each other to form an impromptu community. Based on Java technology as shown in Fig.1 [1] [2], Jini technology uses Java Remote Method Invocation protocols to move code around the network. Network services run on top of the Jini software architecture.

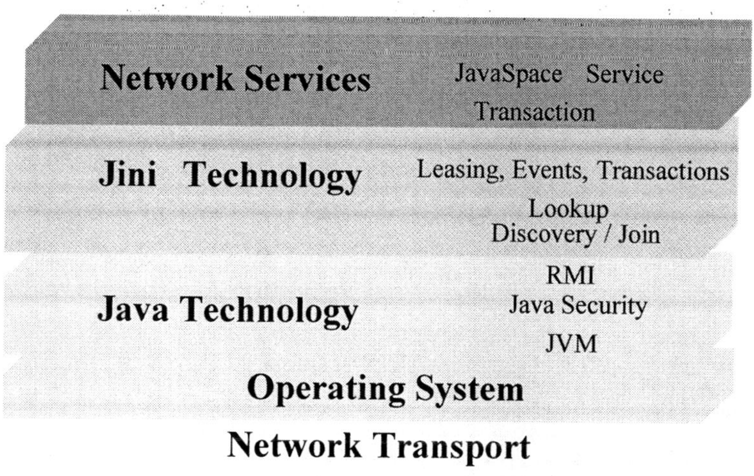

Fig.1 Architecture of Jini Connection Technology

2.1 Lookup Service

The lookup service can be viewed as a directory service, in that services are found and resolved through it. In a Jini community, services register their proxy objects with a lookup service through discovery and join process, and clients query the lookup service to find out the services they want. Jini uses three related discovery protocols, useful in different situation [3] [4]. *Multicast Request Protocol* is used when an application or service first becomes active, and needs to find lookup services in the vicinity. *Multicast Announcement Protocol* is used by lookup services to announce their presence to the services that may have interest in the community. *Unicast Discovery Protocol* is used to establish communications with a specific lookup service known to it in priori over a wide-area network.

But a Jini lookup service does much more than a simple name server. Client sees a service as an interface, including methods that the client will invoke to execute the service, along with any other descriptive attributes. The lookup service maps interfaces seen by clients to set of service proxy objects. Client downloads the service proxy, which is actually RMI stub that can communicate back with the server. This proxy object enables client to use the service without knowing anything about it. Hence, no need for device driver scenario. Although service proxy object is typical scenario of service invocation, i.e. accessing services through RMI method invocation, the downloaded service object can be the service itself or a smart object capable of speaking any private communication protocol.

2.2 Leasing

Access to services in the Jini system is granted on lease basis: A service is requested for a time period and, then, granted for negotiated period between the service user and provider. This lease must be renewed before its expiration. Otherwise, the resources associated with the services are released. For the example, the lookup service grants lease to a service registration and the service should continue to renew the lease. A device can leave the community or fail abruptly without having a chance to deregister itself. So, It is the leasing that enables the Jini system to be kept robust and maintenance-free.

2.3 Remote Events and Transactions

Besides the basic service discovery/join and lookup mechanism, Jini supports *remote events* and *transactions* that help programmers write distributed programs in reliable and scalable fashion. Remote event enables an object to be notified when desired change occurs in the system. These events can be triggered by newly-published services or some state changes of services. For example, a Jini palmtop that registered its interest in printers can be notified by the lookup service, when a printer becomes available. Also, Jini supports two-phase commit (2PC) protocol. By nature, Jini is used to build distributed systems where reliability and robustness are likely to get impaired by partial failures and recovery. But Jini 2PC allows flexibility, in that it does not dictate this protocol to be followed strictly. Rather, it is being left to applications (transaction participants) to implement necessary actions intended by the application logic.

3 Universal Plug and Play

Universal Plug and Play (UPnP) is an architecture for pervasive peer-to-peer network connectivity of intelligent appliances, wireless devices, and PCs of all form factors. Although it's introduced as an extension to the plug and play peripheral model, UPnP is more than a simple extension to it. In UPnP, a device can dynamically join a network, obtain an IP address, convey its capabilities upon request, and learn about the presence and capabilities of other devices. Finally, a device can leave a network smoothly and automatically without leaving any unwanted state behind [6]. Universal Plug and Play leverages TCP/IP and the Web technologies, including IP, TCP, UDP, HTTP and XML, to enable seamless proximity networking in addition to control and data transfer among networked devices in the home and office.

UPnP uses Simple Service Discovery Protocol (SSDP) [7] for service discovery. This protocol is used for announcing a device's presence to others as well as discovering other devices or services. Therefore, SSDP is analogous to the trio of protocols in Jini: discovery, join, and lookup. SSDP uses HTTP over multicast and unicast UDP which are referred to as HTTPMU and HTTPU, respectively.

A joining device sends out a *advertisement (ssdp:alive)* multicast message to advertise its services to control points. Control points have a very similar functionality to Jini's lookup services. A control point, if present, can record it, or other devices may directly see this multicast message as well. In contrast to Jini, UPnP can work with or without the lookup service (control point). The other message of SSDP is *search (ssdp:discover)* multicast message sent when a new control point is added to a network. Any device that hears this multicast should respond to it with a unicast response message.

XML is used to describe device features and capabilities. The aforementioned advertisement message contains a URL that points to an XML file in the network, describing the UPnP device's capability. Hence other devices, by retrieving this XML file, can inspect the features of this device and decide whether it fits their purposes. This XML description allows complex, powerful description of device capability as opposed to Jini's simple service attribute.

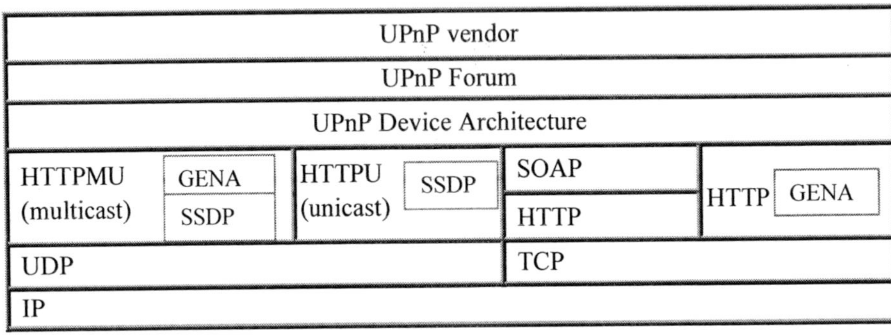

Fig.2 UPnP Protocol Stack

For communication between devices, UPnP uses the protocol stack shown in Fig.2 [6]. According to the latest specification [6], UPnP features can be epitomized as the following five steps.

1. *Discovery*: The UPnP discovery protocol is based on SSDP. When a device added to the network, the device advertises its services to the control points on the network. Similarly, when a control point is added to the network, the UPnP allows that control point to search for devices of interest on the network. The fundamental exchange in both cases is a discovery message containing a few, essential specifics about the device or one of its services, e.g., its type, identifier, and a pointer to more detailed information.

2. *Description:* After a control point has discovered a device, the control point still knows very little about the device. For the control point to learn more about the device and its capabilities, or to interact with the device, the control point must retrieve the device's description from the URL provided by the device in the discovery message. The UPnP description for a device is expressed in XML and includes a list of any embedded devices or service, as well as URLs for control, eventing, and presentation.

3. *Control:* After a control point has retrieved a description of the device, the control point can send actions to a device's service. To do this, a control point sends a suitable control message to the control URL for the service. Control messages are also expressed in XML using the Simple Object Access Protocol (SOAP). Like function calls, in response to the control message, the service returns any action-specific values.

4. *Eventing:* An UPnP description for a service includes a list of actions the service responds to and a list of variables that model the state of the service at run time. The service publishes updates when these variables change, and a control point may subscribe to receive this information. The service publishes updates by sending event messages. Event messages contain the names of one of more state variables and the current value of those variables. These messages are also expressed in XML and formatted using the General Event Notification Architecture (GENA).

5. *Presentation:* If a device has a URL for presentation, then the control point can retrieve a page from this URL, load the page into a browser, and depending on the capabilities of the page, allow a user to control the device and/or view device status.

Another important feature of UPnP is automatic configuration of IP addresses being plugged in. Being introduced for this purpose, AutoIP [13] enables a device to join the network without any explicit administration. When a device is connected to the network, it tries to acquire an IP address from a DHCP server on the network. But in the absence of a DHCP server, an IP address is claimed automatically from a reserved range for the local network use. So, named as AutoIP. The device claims an address by randomly choosing an address in the reserved range and then making an ARP request to see if anyone else has already claimed that address.

4 Salutation

Salutation is another major cooperation architecture, which is being developed by the Salutation Consortium, to solve the problems of service discovery and utilization among a broad set of appliances and equipment and in an environment of widespread connectivity and mobility. Given the diverse nature of target devices, it is processor, operating system and communication protocol independent. The architecture provides a standard method for applications, services and devices to describe and to advertise their capabilities to other applications, services and devices. The architecture also enables application, services and devices to search other applications, services or devices for a particular capability, and to request and establish interoperable sessions with them to utilize their capabilities [8] [9].

As shown in Fig.3 [8], the Salutation architecture is composed of two major components: *Salutation Manager* and *Transport Manager*. The Salutation Manager is the core of the architecture, similar to the lookup service in Jini or the control point in UPnP. It is defined as a service broker: "A service provider registers its capability with a Salutation Manager. When a client ask its local Salutation Manager for a service search, the search is performed by coordination among Salutation Managers. Then, the client can use the returned service." A Salutation Manager sits on the Transport Managers that provide reliable communication channels, regardless of what the underlying network transports are.

The Salutation Manager provides a transport-independent interface to Server and Client applications. This interface (SLM-API) includes service registration, service discovery, and service access function. The communication protocol independence of Salutation architecture is achieved by the interface (SLM-TI) between Salutation Manager and Transport Manager. Transport Manager is an entity, dependent on the network transport it supports. A Salutation Manager may have more than one Transport Manager, in case it is attached to multiple, physically different networks. But Salutation Manager sees its underlying transport through the transport-independent interface (SLM-TI).

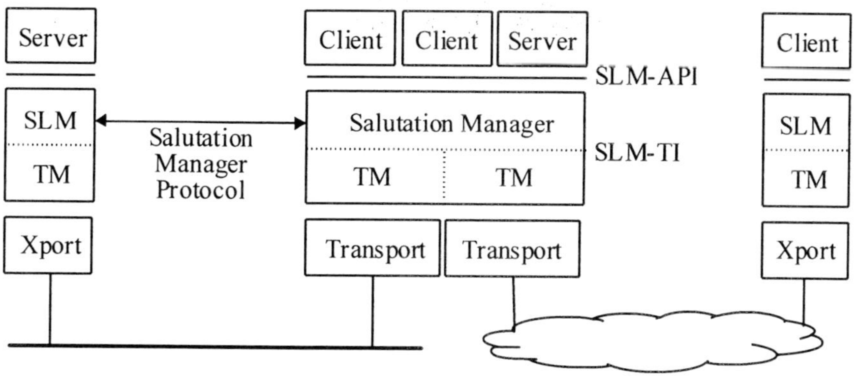

Fig.3 Model of the Salutation Manager

The main tasks provided by the Salutation Manager can be summarized as follows.

- **Service Registry**: The Salutation Manager contains a registry to keep information about Services. A client registers or unregisters itself. All registration is done with the local Salutation Manager or near one connected to the client. This is correspondent to the lookup service in Jini.

- **Service Discovery**: The Salutation Manager discovers other Salutation Managers and services registered there. Remote services are discovered by matching type(s) and set of attributes specified by local Salutation Manager. This communication protocol between Salutation Managers is called the Salutation Manger Protocol using Sun's ONC RPC. This unique feature, called capability exchange, is needed, because services are basically registered with the local Salutation Manager in the same equipment. This cooperation among Salutation Managers forms the conceptually same lookup service, but distributed over the network, as Jini does.

- **Service Availability**: A client application can ask the local Salutation Manager to periodically check the availability of services. This checking is done between the local manager and the corresponding manager. This is a narrow version of Jini's Remote Event concept.

- **Service Session Management**: This session management addresses the service invocation aspect of Salutation. A service session is established when a client wants to use a service discovered through Service Discovery. The service session is operated in one of 3 different modes: *native mode, emulated mode, and salutation mode*. The Salutation Manager may or may not be involved in message exchanges in the service session, depending on the modes. In the *native mode*, messages are exchanged through a native protocol and Salutation Manager is never involved in message exchange. In the *emulated mode*, the Salutation Manager Protocol is used to carry messages between client and service but Salutation Manager doesn't inspect the contents. In the *salutation mode*, Salutation Managers not only carry messages, but also define the message formats to be used in the session.

A *Functional Unit* is defined as a basic building block in Salutation architecture. In other words, it is the minimal meaningful function to constitute a client or service. A collection of Functional Units defines a Service Record. For example, a fax service can be defined by [Print], [Scan], and [Fax Data Send] Functional Units. Each functional unit is composed of descriptive attribute record. These Service/Funtional Unit/Attribute records are specified with ISO 8824 ASN.1.

Salutation-Lite [10] is also worth to mention here. Salutation-Lite is a scaled down version of the Salutation architecture targeted at devices with small footprints. The Salutation Consortium envisions that Salutation-Lite has tremendous applicability to small information appliances such as palm-size and hand-held computers (i.e. Palm and WinCE devices). Salutation-Lite also lends itself well to low bandwidth network such as IR and Bluetooth.

5 SERVICE LOCATION PROTOCOL AND OTHERS

Service Location Discovery (SLP) [17] is an IETF version of service discovery protocol but it has unique backgrounds, target areas, and features, as other service discovery protocols do. SLP is a decentralized, lightweight, scale and extensible protocol for service discovery within a site [16]. SLP defines *Service URL* which defines service type and address for the service. For example, "service:printer:lpr://hostname" is the Service URL for line printer service available at hostname. Based this Service URL, a user browses services available in its site and makes use of selected services to meet the user's need. For example, a user (application) uses SLP to find out any color printer on the same floor.

There are three agents in SLP: *User Agent (UA), Service Agent (SA), and Directory Agent (DA)*. UA is a software entity that sends service discovery requests on behalf of a user application. SA is an entity that advertises service on behalf of a service. As a centralized service information repository, DA caches advertisements from SAs and, afterwards, responds to requests from UAs. An SA advertises itself by registering with a DA. This registration message contains the URL for the advertised service, lifetime for the service, and a set of descriptive attributes for the service. The SA should periodically refresh the registration with DA before its expiration. This lifetime is meant to prevent the network from being left in transient state and similar concept is found at other service discovery protocols such as Jini and UpnP. A DA caches the registration and sends an acknowledge message to the SA. A UA send a service request message to the DA to request the location of a service. Then, the DA responds with a service reply message including the URLs of the services matched against the UA needs. Now, the UA can access the service pointed by the returned URL. In SLP, DA is optional. There may be no DA in small networks. In this case the UA's service request message is directly sent to SAs.

SLP supports service browsing and string-based query for service attributes which allow UA to select the most appropriate service from among services in the network. The UA can request query operators such as AND, OR, comparators (=, <, <=, >, >=), and substring matching. This is more powerful than others. For example, in Jini, service attribute matching can be done only against equality.

Finally, SLP is said to be a solution to the intranet service discovery needs but it scales well to larger network. The scalability is supported by various features such as the minimal use of multicast messages, *scope* concept, and multiple DAs.

Bluetooth protocol stack also contains a SDP [14] for service discovery. Since Bluetooth SDP is designed specically for Bluetooth environments, it supports limited functionality, compared to other service discovery protocols. Basically, SDP supports search by service class, search by service attributes, and service browsing. Service browsing is used when a client has no priori knowledge about services available in the client's vicinity. Service discovery application profile [15] defines protocols and procedures used by a service discovery application to locate services in other devices. Bluetooth SDP runs on a predefined connection-oriented channel of L2CAP.

Bluetooth SDP is optimized for Bluetooth devices with limited complexity. Thus, it addresses primarily service discovery problem. It provides neither access to services, brokering of service, service advertisement, nor service registration. There's no event notification when services become unavailable. Therefore, other service discovery protocol might be used to complement these lacks. For example, Salutation can be used above Bluetooth SDP. Such mapping [11] seems to be neat because of Salutation's transport-independent architecture.

There are other players in this area: Zero Configuration Networking (zeroconf) [20], MIT's INS (Intentional Naming System) [21], and the Berkeley Service Discovery Service [22]. With a different objective, each of them takes a different approach from others. As a result, they have some strong and weak features, relatively compared to other protocols.

6 COMPARISON

Several service discovery protocols are proposed to facilitate dynamic cooperation among devices/services with minimal administration and human intervention. In order to be able to support the impromptu community, they should provide the means to announce its presence to the network, to discover services in the neighborhood, and to access to services. Basically, all Jini,

UPnP, Salutation, and SLP address these aspects, but in different perspectives. A direct comparison must be avoided, since they put different weights on the above functionality. Nevertheless, such comparison is tried here, since it would be helpful to understand each of them. Table 1 summarizes the features of major service discovery protocols.

Jini and UPnP envision pervasive computing environments being enabled by their solutions, whereas Salutation and SLP are primarily dealing with the service discovery problem. Note that Jini provides 2PC transaction and JavaSpace to help develop network services [3]. UPnP's SSDP is just a part of UPnP specification. A good comparison among Jini, UPnP, and Salutation is presented in [19].

Jini has a dependence on Java to enable all its promises. It assumes that devices support Java Virtual Machine, even though a Jini-proxy can be used for a cluster of resource-poor devices [3]. Moreover, Jini/RMI is not supported by J2ME CLDC (Connected Limited, Device Configuration) configuration for small information devices such as cell phone, pager, and POS [24].

Jini's service proxy concept is one of strongest features not found at others. But this no-need-for-drivers scenario presumes the Jini devices to support standard interfaces are already available in the network. It's not as simple as it sounds, since it means all manufactures of a certain device type must consent to the standard interface. First, the standardizations for printer and storage device interfaces are under way by the consortia of manufactures.

UPnP relies on the existing IP and Web technology. It seems unique in terms of its XML use for service/device description. XML allows for powerful description of device capability, control command issued to the device, event from it. UPnP introduces new features for self-configuration which exploit AutoIP and DHCP, but these features are also found in IPv6 [19]

The Salutation is well defined but confined to the service discovery protocol and session management. Salutation accordingly doesn't address features like remote event notification, which are no doubt useful in distributed environment. When it comes to transport protocol, IP is given top priority by Jini, UPnP/SSDP, and SLP. Salutation can operate in any network, including IP, IR and, in the future, wireless. This transport independence is the strongest feature of Salutation.

More than one SLP DAs are likely to be deployed for an enterprise network, since a DA becomes the single point of failure. These DAs can be organized in a hierarchy to provide better performance. Also, there may be some overlap in their coverage of organization/departments to provide reliability. This interaction or cooperation between DAs for performance and reliability is being explored by SLP society. SLPv2 can ensure the integrity and authenticity of SLP messages by including authentication information in SLP message. It deals with security problem directly, while others have to rely on other security protocols.

7 CONCLUSION

This survey attempts to take a look at key service discovery protocols that address more or less the same concepts. These service discovery protocols are vying with one another to be final winners. On the other hand, they are also making efforts to integrate themselves with other protocols. A Jini-to-SLP bridge developed by Sun is one of such efforts. Salutation architecture has already aligned its architecture with SLP (enhanced to support directory-based service discovery by utilizing SLP). As another example, the use of Jini's service proxy object in Salutation platform is made possible by [DOC Storage] Functional Unit, although its use is not limited to Jini proxy object. It can be used for device drivers and application programs.

There seems to be no clear choice today. All of them have wonderful architectures but technological superiority is not the only factor to be prevalent in the market. We expect more changes and competitions in the upcoming years, accelerated by industry feedback following their deployment. It would be very interesting to watch where this dynamic evolution converges.

	Jini	UPnP	Salutation	SLP
Home Page	www.sun.com/jini	www.upnp.org	www.salutation.org	www.svrloc.org
Main Entities	Lookup Service, Client , Service	Control Point, Devices (Services)	Salutation Manager, Transport Manager, Client, Server	Directory Agent, Service Agent, User Agent
Service Repository	Lookup Service	Control Point	A set of SLMs (Salutation Managers)	DA (Directory Agent)
Service Announcement	Discovery/Join protocol	Advertisement (ssdp:alive)	Registering with local Salutation Manager	Service registration
Service Discovery	Query to lookup service	Contact control point, or listen to advertisement	Query to local SLM and cooperation among SLMs	Contact DA, or multicast to SAs
Access to Service	Service proxy object based on RMI	Invoking action to the service (SOAP), Query for variable state	Service Session management	Service type (service protocol) for the discovered service
Service Description and Scoping	Interface type and attribute matching	Description in XML	FU (Functional Unit) and attributes within it	Service type And attribute matching (fairly powerful matching)
Service Registration Lifetime	Leasing	CACHE-CONTROL header in alive message	No	Lifetime in service registration
Service Group	Group	No	No	Scope
Event Notification	Remote Events	Service publishes event(GENA) when state variable changes	Availability Checking (periodic & automatic)	No
Others	Java-centric architecture	Automatic con-figuration (AutoIP)	Transport independence	Authentication security feature

Table 1 Summary of major service discovery protocol

REFERENCES

[1] Sun Microsystems, "Jini Connection Technology," http://www.sun.com/jini

[2] Sun Microsystems, "Jini Community Resources: Jini Technology Architectural Overview," January 1999. http://www.sun.com/jini/whitepapers/architecture.html

[3] Sun Microsystems, "Jini Community Resources: Jini Specification v1.0.1," www.sun.com/jini/specs

[4] W. Keith Edwards, *Core JINI, The Sun Microsystems Press Java Series*, Prentice Hall, 1999.

[5] Microsoft Corporation, "Universal Plug and Play: Background," http://www.upnp.org/resources/UpnPbkgnd.htm

[6] Microsoft Corporation, "Universal Plug and Play Device Architecture Version 1.0," June 8, 2000. http://www.upnp.org/UpnPDevice_Architecture_1.0.htm

[7] Yaron Y. Goland, Ting Cai, Paul Leach, Ye Gu, and Shivaun Albright, "Simple Service Discovery Protocol," IETF Draft draft-cai-ssdp-v1-03.txt, October 28, 1999. http://www.ietf.org/internet-drafts/draft-cai-ssdp-v1-03.txt

[8] Salutation Consortium, "Salutation Architecture Specification Version 2.0c – Part 1," The Salutation Consortium, June 1, 1999. http://www.salutation.org

[9] Salutation Consortium, "Salutation Architecture Specification Version 2.0c – Part 2," The Salutation Consorium, June 1, 1999. http://www.salutation.org

[10] Bob Pascoe, "Salutation-Lite: Find-and-Bind Technologies for Mobile Devices," Salutation Consortium, June 6, 1999. http://www.salutation.org/whitepaper/Sal-Lite.PDF

[11] Brent Miller, "Mapping Salutation Architecture APIs to Bluetooth Service Discovery Layer," Bluetooth Consortium 1.C.118/1.0, July 1, 1999.

[12] Bob Pascoe, "Salutation Architectures and the newly defined service discovery protocols from Microsoft and Sun," Salutation Consortium, June 6, 1999. http://www.salutation.org/whitepaper/Jini-UPnP.PDF

[13] Ryan Troll, "Automatically Choosing an IP Address in an Ad-Hoc IPv4 Network," IETF Draft draft-ietf-dhc-ipv4-autoconfig-05.txt, March 2, 2000. http://www.ietf.org/internet-drafts/draft-ietf-dhc-ipv4-autoconfig-05.txt

[14] Bluetooth Consortium, "Specification of the Bluetooth System Core Version 1.0 B: Part E, Service Discovery Protocol (SDP)," Nov 29, 1999. http://www.bluetooth.com/developer/specification/specification.asp

[15] Bluetooth Consortium, "Specification of the Bluetooth System Profiles Version 1.0 B: Part K:2, Service Discovery Application Profile," Dec 1, 1999.
http://www.bluetooth.com/developer/specification/specification.asp

[16] IETF SVRLOC Working Group, "Service Location Protocol Home Page," http://www.srvloc.org

[17] E. Guttman, C. Perkins, J. Veizades, and M. Day, "Service Location Protocol, Version 2," IETF RFC 2608, June 1999. http://www.ietf.org/rfc/rfc2608.txt

[18] E. Guttman, C. Perkins, and J. Kempf, "Service Templates and Service: Schemes," IETF RFC 2609, June 1999. http://www.ietf.org/rfc/rfc2609.txt

[19] Rekesh John, "UPnP, Jini and Salutation – A look at some popular coordination frameworks for future networked devices," California Software Labs, June 17, 1999. http://www.cswl.com/whiteppr/tech/upnp.html

[20] IETF ZEROCONF Working Group, "Zero Configuration Networking (zeroconf)," http://www.ietf.org/html.charters/zeroconf-charter.html

[21] INS: Intentional Naming System, http://wind.lcs.mit.edu/projects/ins

[22] The Berkeley Service Discovery Service, http://www.cs.berkeley.edu/~caerwin/sds-project.html

[23] R. Droms, "Dynamic Host Configuration Protocol," IETF RFC 2131, March 1997. http://www.ietf.org/rfc/rfc2131.txt

[24] Sun Microsystems, "Java 2 Platform Midcro Edition (J2ME) Technology for Creating Mobile Devices," White Paper, May 19, 2000. http://java.sun.com/products/cldc/wp/KVMwp.pdf

[25] World Wide Web Consortium, "Extensible Markup Language (XML)", http://www.w3.org/XML

Chapter 2

REDUCING PCS SIGNALING TRAFFIC THROUGH DELAY TECHNIQUES

Yi-Bing Lin
Department of Computer Science and Information Engineering
National Chiao Tung University
TAIWAN, R.O.C.
liny@csie.nctu.edu.tw
and
Jyhi-Kong Wey
Telecommunication Laboratories
Chunghwa Telecom Co., Ltd
TAIWAN, R.O.C.
E-mail: jkwey@ms6.hinet.net

Abstract: The mobility management messages generated by Personal Communications Services may result in heavy signaling traffic to the telephone network. This paper describes two delay registration approaches, MS-controlled delay registration (MS-DR) and VLR-controlled delay registration (VLR-DR), to reduce the registration traffic. Analytic models are proposed to investigate the performance of MS-DR and VLR-DR. Our study indicates that following appropriate guidelines, these delay registration approaches effectively reduce the number of messages at the cost of losing very few number of call deliveries to a mobile station. Both MS-DR and VLR-DR can be easily implemented without modifying the existing roaming management protocols such as IS-41 and GSM MAP.

Key-Words: - Delay registration, Home Location Register, location tracking, mobility management, Visitor Location Register

1 INTRODUCTION

Cellular telephony provides voice and data services to the users with mobility. To deliver services to the mobile users, the cellular network is capable of tracking the locations of the *mobile stations* (MSs). These capabilities are achieved by the roaming management protocols such as IS-41 [3] and GSM MAP [4]. Two basic operations in roaming management are *registration* (the process that an MS informs the system of its current location) and *location tracking* (the process that the system locates the MS). Location tracking is required when the network attempts to delivery a call to the MS. Details of mobility management can be found in [9]. Following the standard PCS registration procedures, an MS initiates the registration procedure when it enters a location area (LA), and the mobility databases (VLR and HLR) are modified to store the current location (the address of the LA) of the MS. The registration messages created by the mobile phone systems may result in heavy signaling traffic to the public switched telephone network (PSTN). Thus, it is desirable to reduce the network overhead caused by location update. Several studies focused on reducing the location update overhead (see [1, 5] and the references therein). Based on the concept of *delay registration*, this paper proposes two simple strategies to reduce the location update traffic. The strategies are described below:

MS-Controlled Delay Registration: It is possible that the MS only resides in an LA for a short period and no phone call arrives before the MS leaves the LA. In this case, the registration for entering the LA is not desirable. In *MS-controlled delay registration* (MS-DR), an MS does not issue a registration request when it enters a new LA. Instead, the MS waits for a period (called the *registration delay period*), and if it is still in the LA after the period expires, then a registration request is actually issued. MS-DR can effectively eliminate the registration traffic for MSs with short LA residence times. However, call deliveries to the MS may arrive before the registration delay period expires (and before the MS leaves the LA). In this case, these calls will be lost. Thus, the length of the registration delay period should be carefully selected to avoid lost calls. To implement MS-DR, we only need to make minor modifications to the MS, and no modifications to VLR/HLR are required. In modern MS design, the modifications can be done by remotely downloading a small piece of code from the mobile network to the MS through the radio link. The user will not notice the modification. Also, MSs with MS-DR can co-exist with the traditional MSs (that do not implement MS-DR) without affecting their operations. MS-DR will be modeled in Section 2.

VLR-Controlled Delay Registration: In *VLR-controlled delay registration* (VLR-DR), the VLR may collect several (say, N) registration requests before it issues a registration notification message to the HLR (in SS7, a TCAP message can deliver several registration requests within the message [10]). In this case, $N-1$ registration messages can be saved. Similar to MS-DR, call deliveries to an MS may be lost during the MS's registration delay period. VLR-DR can be implemented by adding the capability of processing multiple-component SS7 messages in VLR/HLR (this feature already exists in SS7 TCAP [2]). No change to the MS micro kernel and the mobility management protocols (such as IS-41 and GSM MAP) is required. No modification to MS is required. Traditional VLRs (that do not support VLR-DR) can co-exist with the HLR/VLR that exercise VLR-DR. VLR-DR will be modeled in Section 3.

In the study of MS-DR and VLR-DR, two output measures are considered:

The portion of saved registration messages. Define registration delay period as the period before an MSmoves in an LA LA_1 and when the registration is performed. If the MS moves out of LA_1 before the registration delay period expires, then the registration operation to LA_1 is saved. Thus, in MS-DR, the portion β_{MS} of saved registration messages is defined as the probability that an MS does not issue a registration after it enters an LA.

In VLR-DR, an MS always issues a registration request when it enters an LA. Without loss of generality, we assume that every VLR covers exactly one LA. In VLR-DR, if VLR collects N registration requests before it issues a registration notification message to the HLR, then the portion β_{VLR} of saved registration messages is $\frac{N-1}{N}$.

The no-lost-call probability. During the registration delay period, call deliveries to the MS may occur. If so, these calls are lost, and the service quality is degraded. The no-lost-call probability α_{MS} for MS-DR (and α_{VLR} for VLR-DR) is defined as the probability that no call delivery to the MS occurs during the registration delay period. The larger the no-lost-call probability α_{MS} (α_{VLR}), the better the performance of MS-DR (VLR-DR).

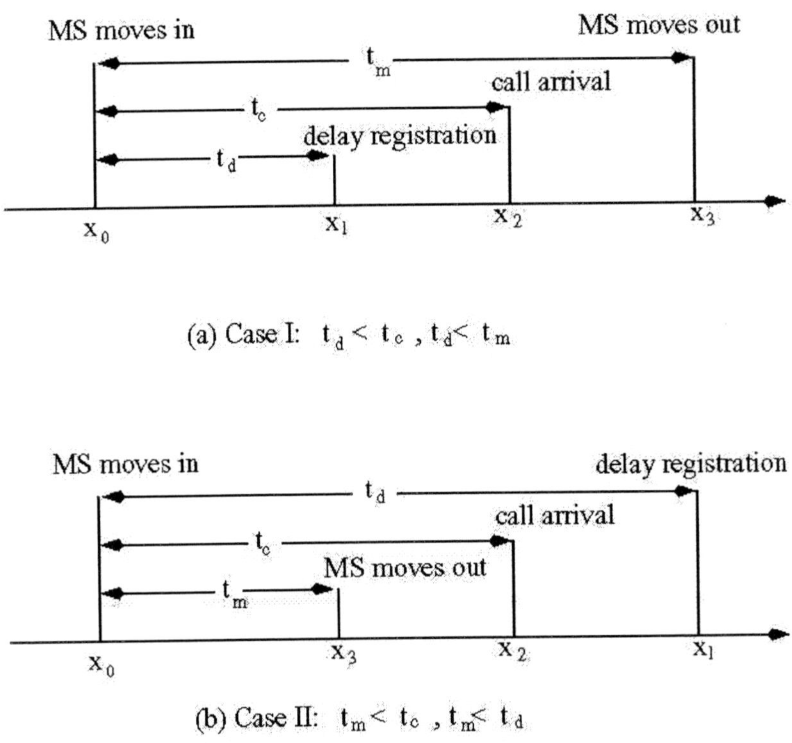

Figure 1: The Timing Diagram

This paper studies how the registration delay period affects the no-lost-call probability α_{MS} (α_{VLR}) and the portion β_{MS} (β_{VLR}) of saved messages. Our study will provide guidelines to select the registration delay period so that satisfactory performance can be achieved.

2 MS-Controlled Delay Registration

This section investigates the performance of MS-DR. Let t_d be the registration delay period. Consider Fig. 1 (a) where an MS moves in an LA at time x_0 and moves out of the LA at time x_3; i.e., the LA residence time is $t_m = x_3 - x_0$. The registration is performed at time x_1 (i.e., the registration delay period is $t_d = x_1 - x_0$. The first call arrival to the MS (after the MS enters the LA) occurs at time x_2 and the period between the MS entrance of the LA and the next call arrival is $t_c = x_2 - x_0$. Suppose that the LA residence time t_m has the density function $f_m(t_m)$, and the distribution function $F_m(t_m)$. If the call arrivals to the MS form a Poisson process with rate λ, then from the memoryless property of the exponential distribution, t_c also has the same exponential density function. We derive the no-lost-call probability α_{MS} and the portion β_{MS} of saved registration messages by considering the fixed registration delay period and the exponential registration delay periods.

2.1 Fixed Registration Delay Period

Suppose that the registration delay period t_d is a constant, then the no-lost-call probability $\alpha_{MS,f}$ can be derived as follows. It is clear that the next call arrival (after the MS moves in the LA) is not lost if $t_c > \min(t_d, t_m)$ (see Fig. 1 (a) and (b)). In other words,

$$\begin{aligned}\alpha_{MS,f} &= P_r[t_c > \min(t_d, t_m)] \\ &= P_r[t_d < t_c, t_d < t_m] + P_r[t_m < t_c, t_m < t_d] \\ &= e^{-\lambda t_d}[1 - F_m(t_d)] + \int_{t_m=0}^{t_d} f_m(t_m) e^{-\lambda t_m} dt_m \qquad (1)\end{aligned}$$

A registration operation is saved if the MS leaves the LA before the registration delay time expires. Thus,

$$\beta_{MS,f} = \Pr[t_d > t_m] + \int_{t_m=0}^{t_d} f_m(t_m)dt_m = F_m(t_d) \quad (2)$$

To derive $\alpha_{MS,f}$ and $\beta_{MS,f}$, we consider two LA residence time distributions (IPP and Gamma). The IPP distribution is appropriate to represent the user moving pattern with the short and long LA residence time categories (to be elaborated). The Gamma distribution is appropriate for general moving patterns. It was shown [7] that the Gamma distribution can be used to approximate many distributions as well as measured data from PCS field trials.

IPP LA Residence Times: Suppose that the LA residence times of a mobile user can be categorized into two groups: short residence time group and long residence time group. For example, a mobile user may drive from home to the work place by traveling several towns (and thus several LAs). During this period, the LA residence times are short. The user may stay at the work place for a long period (say, 10 hours) and then moves back home again. During the working hours, the user may reside in an LA for a long period. Suppose that the long and the short LA residence times are represented by exponential distributions with means $1/\eta_1$ and $1/\eta_2$, respectively, where θ portion of the LA residence times belong to the short category and $1-\theta$ of them are long LA residence times. Then the mobile user's LA residence time distribution can be represented by an *interrupted Poisson process* (IPP) [8] with the density function

$$f_m(t_m) = \theta\eta_1 e^{-\eta_1 t_m} + (1-\theta)\eta_2 e^{-\eta_2 t_m} \quad (3)$$

and the distribution

$$F_m(t_m) = 1 - \theta e^{-\eta_1 t_m} - (1-\theta)e^{-\eta_2 t_m}$$

With the IPP LA residence times, the no-lost-call probability is denoted as $\alpha_{MS,f,IPP}$ and the portion of saved registration messages is denoted as $\beta_{MS,f,IPP}$. Thus, (1) is re-written as

$$\begin{aligned}\alpha_{MS,f,IPP} &= e^{-\lambda t_d}\left[\theta e^{-\eta_1 t_d} + (1-\theta)e^{-\eta_2 t_d}\right] \\ &+ \int_{t_m=0}^{t_d}\left[\theta\eta_1 e^{-\eta_1 t_m} + (1-\theta)\eta_2 e^{-\eta_2 t_m}\right] e^{-\lambda t_m} dt_m \\ &= \left(\frac{\theta}{\eta_1+\lambda}\right)\left[\eta_1 + \lambda e^{-(\eta_1+\lambda)t_d}\right] + \left(\frac{1-\theta}{\eta_2+\lambda}\right)\left[\eta_2 + \lambda e^{-(\eta_2+\lambda)t_d}\right]\end{aligned} \tag{4}$$

and (2) is re-written as

$$\beta_{MS,f,IPP} = 1 - \theta e^{-\eta_1 t_d} - (1-\theta)e^{-\eta_2 t_d} \tag{5}$$

Gamma LA Residence Time: Suppose that the LA residence times of an MS have a Gamma distribution with the scale parameter η and the shape parameter γ (i.e., the mean is γ/η and the variance is γ/η^2). Then

$$f_m(t_m) = f_{m,G}(t_m, \eta, \gamma) = \frac{\eta^\gamma t^{\gamma-1} e^{-\eta t}}{\Gamma(\gamma)} \tag{6}$$

(where $\Gamma(\bullet)$ is the Gamma function) and its Laplace transform is

$$f_m^*(s) = \left(\frac{\eta}{\eta+s}\right)^\gamma \tag{7}$$

Denote the Gamma LA residence time distribution as $F_{m,G}(t_m, \eta, \gamma)$. From (6), (1) is re-written as

$$\alpha_{MS,f,G} = e^{-\lambda t_d}\left[1 - F_{m,G}(t_d, \eta, \gamma)\right] + \left(\frac{\eta}{\eta+\lambda}\right)^{\gamma} F_{m,G}(t_d, \eta+\lambda, \gamma) \tag{8}$$

From (2),

$$\beta_{MS,f,G} = F_{m,G}(t_d) \tag{9}$$

In (8) and (9), the Gamma distribution function Fm,G(•). cannot be computed directly. An inversion technique for the Laplace Transform [6] can be used to generate the approximate sequence to calculate Fm(t):

$$F_m(t) = \lim_{n\to\infty}\left[\frac{(-1)^n s^{n+1}}{n!}\right]\left\{\frac{d^n\left[\frac{f_m^*(s)}{s}\right]}{ds^n}\Bigg|_{s=\frac{n+1}{t}}\right\}$$

where $d^n\left[\frac{f_m^*(s)}{s}\right]/ds^n$ can be computed iteratively [11] as

$$\frac{d^n\left[\frac{f_m^*(s)}{s}\right]}{ds^n} = \begin{cases} f_m^*(s)/s, & n = 0 \\ \frac{1}{s}\left\{\frac{d^n f_m^*(s)}{s^n} - \frac{n d^{n-1}\left[\frac{f_m^*(s)}{s}\right]}{ds^{n-1}}\right\}, & n > 0 \end{cases}$$

2.2 Exponential Registration Delay Time

If t_d is exponentially distributed with mean $1/\mu$, then the no-lost-call probability $\alpha_{MS,e}$ is expressed as

$$\begin{aligned}\alpha_{MS,e} &= \int_{t_m=0}^{\infty}\int_{t_d=0}^{t_m}\int_{t_c=t_d}^{\infty} \lambda e^{-\lambda t_c}\mu e^{-\mu t_d} f_m(t_m)dt_m dt_d dt_c \\ &+ \int_{t_m=0}^{\infty}\int_{t_d=t_m}^{\infty}\int_{t_c=t_d}^{\infty} \lambda e^{-\lambda t_c}\mu e^{-\mu t_d} f_m(t_m)dt_m dt_d dt_c \\ &= \frac{\mu}{\mu+\lambda} + \frac{\lambda f_m^*(\lambda+\mu)}{\mu+\lambda}\end{aligned} \tag{10}$$

and the portion $\beta_{MS,e}$ of saved registration messages is expressed as

$$\beta_{MS,e} = \int_{t_d=0}^{\infty}\int_{t_m=0}^{t_d} f_m(t_m)\mu e^{-\mu t_d} dt_m dt_d = f_m^*(\mu) \tag{11}$$

IPP LA Residence Times: If the density function of t_m is expressed in (3), then the Laplace transform of $F_m(t_m)$ is expressed as

$$f_m^*(s) = \frac{\theta\eta_1}{\eta_1+s} + \frac{(1-\theta)\eta_2}{\eta_2+s} \tag{12}$$

From (10), (11) and (12), we have

$$\alpha_{MS,e,IPP} = \frac{\mu}{\lambda+\mu} + \frac{\lambda}{\mu+\lambda}\left[\frac{\theta\eta_1}{\eta_1+\mu+\lambda} + \frac{(1-\theta)\eta_2}{\eta_2+\mu+\lambda}\right] \tag{13}$$

$$\beta_{MS,e,IPP} = \frac{\theta\eta_1}{\eta_1+\mu} + \frac{(1-\theta)\eta_2}{\eta_2+\mu} \tag{14}$$

Gamma LA Residence Time: Suppose that t_m has a Gamma distribution with the scale parameter γ and the shape parameter η. Denote $\alpha_{MS,e,G}(\beta_{MS,e,G})$ as $\alpha_{MS,e}(\beta_{MS,e})$ with Gamma LA residence time. Then (10) and (11) are re-written as

$$\alpha_{MS,e,G} = \frac{\mu}{\mu+\lambda} + \frac{\lambda\eta^\gamma}{(\mu+\lambda)(\eta+\lambda+\mu)^\gamma} \quad (15)$$

$$\beta_{MS,e,G} = \left(\frac{\eta}{\eta+\mu}\right)^\gamma \quad (16)$$

2.3 Numerical Examples

Based on the analytic models in Sections 2.1 and 2.2, we use several numerical examples to study the performance of MS-DR. Fig. 2 plots α_{MS} (the solid curves) and β_{MS} (the dashed curves) against θ with fixed delay registration period t_d. Assume that on the average, a mobile user stays in the work place for 10 hours (i.e., η_1=0.1 per hour). The user drives through several LAs between home and the work place. The average LA residence times for these LAs is 10 minutes (η_2=60η_1). The number of these traveled LAs is controlled by the parameter θ. If θ = 0.1 (θ=0.5), then on the average, the user travels through 9 LAs (1 LA) before arriving at the work place. The arrival rate of the call deliveries to the user is 0.5 calls per hour (λ=5η_1). The following results are observed:

The effect of t_d: It is apparent that a large t_d results in a large $\beta_{MS,f}$ and a small $\alpha_{MS,f}$. When t_d=1/λ, MS-DR effectively reduces the number of registrations by 60%-92% at the cost of low no-lost-call probabilities between 67%-87%. For t_d =(1/16/λ), the no-lost-call probability is around 95%, and the number of registrations is reduced by 27%-48%.

The above observations suggest that a small t_d (e.g., t_d=(1/16λ)) should be selected to save reasonable amount of the registration traffic without degrading the no-lost-call performance.

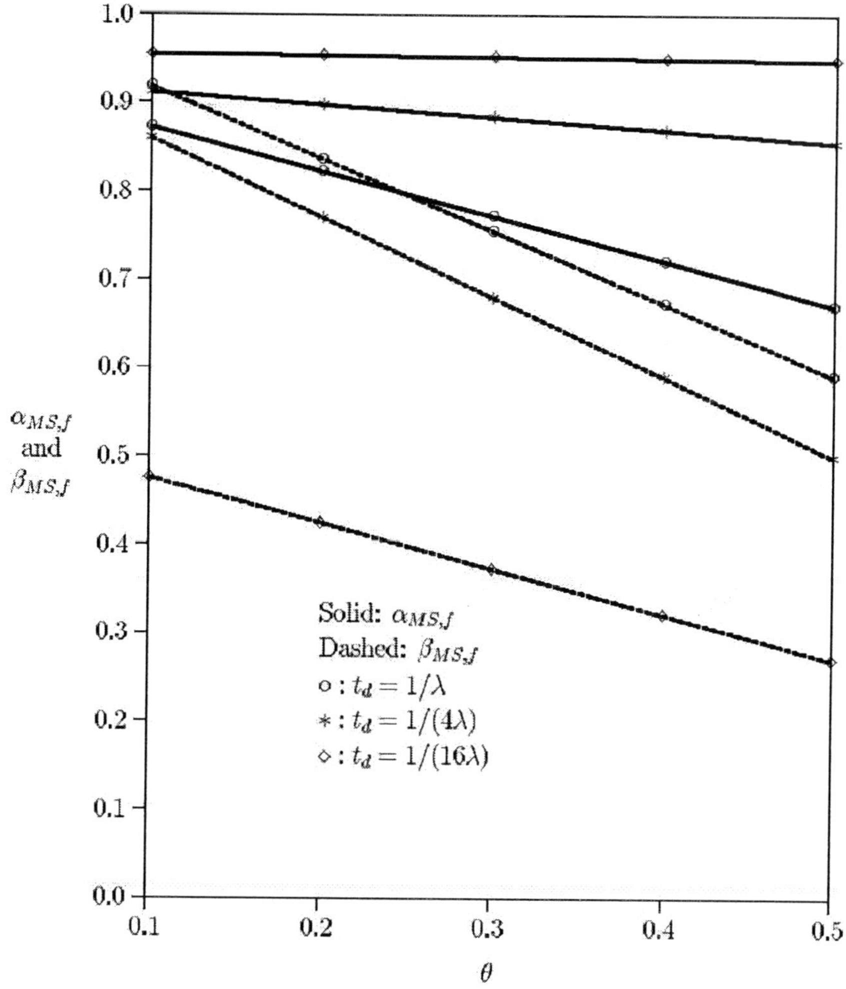

Figure 2: The Performance of MS-DR with Fixed Registration Delay Period and IPP LA Residence Times (η_2=60η_1, λ=5η_1)

The effect of θ: With a large θ, it is more likely to observe large LA residence times. Thus, it is obvious that both $\alpha_{MS,f}$ and $\beta_{MS,f}$ decrease as θ increases. A non-trivial observation is that when t_d is small, $\alpha_{MS,f}$ is insensitive to the change of θ. This observation leads to the conclusion that a small t_d (e.g., $t_d = 1/(16\lambda)$) should be selected to guarantee stable no-lost-call performance that is only slightly affected by the change of θ.

The effect of η_2: Fig. 3 illustrates the effect of η_2 where $\eta_2 = 30\eta_1$ and $60\eta_1$ are considered (i.e., the average short LA residence times are 20 and 10 minutes, respectively). Both $\alpha_{MS,f}$ and $\beta_{MS,f}$ decrease as η_2 decreases (i.e., as the mean of the short LA residence times increases). A non-trivial observation is that $\beta_{MS,f}$ is more sensitive to η_2 than $\alpha_{MS,f}$ is.

The effect of λ: As shown in Fig. 4, the effect of λ is similar to the effect of η_2. In this figure, $\lambda = 5\eta_1$ (0.5 calls per hour) and $10\eta_1$ (one call per hour) are considered.

The effect of the LA residence time distributions: Fig. 5 shows the effect of the IPP and the Gamma distributions. In this figure, for every θ value in (3), the η and the γ values in (6) are derived as

$$\eta = \frac{\theta\eta_2^2 + (1-\theta)\eta_1^2}{\theta\eta_1\eta_2^2 + (1-\theta)\eta_1^2\eta_2} \quad \text{and} \quad \gamma = \frac{[\theta\eta_2 + (1-\theta)\eta_1]^2}{\theta\eta_2^2 + (1-\theta)\eta_1^2}$$

so that the corresponding IPP and the Gamma LA residence time distributions have the same means and variances. The figure indicates that when θ is small, the $\alpha_{MS,f}$ performance for the IPP distribution is better than that for the Gamma LA residence time distribution. The result reverses when θ is large. In all cases, the $\beta_{MS,f}$ performance for the IPP distribution is better than the Gamma distribution. However, the performance for the Gamma distribution is reasonably good. In other words, MS-DR is effective for both the IPP and the Gamma residence time distributions.

The effect of the t_d distribution: Fig. 6 illustrates the effect of the fixed t_d and the exponential t_d. The figure indicates that the α_{MS} performance for the exponential t_d is better than that for the fixed t_d. On the other hand, the β_{MS} performance for the fixed t_d is better than that for the exponential t_d. Thus, if no-lost-call performance is important, and medium β_{MS} performance is acceptable, then exponential t_d should be selected. Otherwise, fixed t_d should be considered.

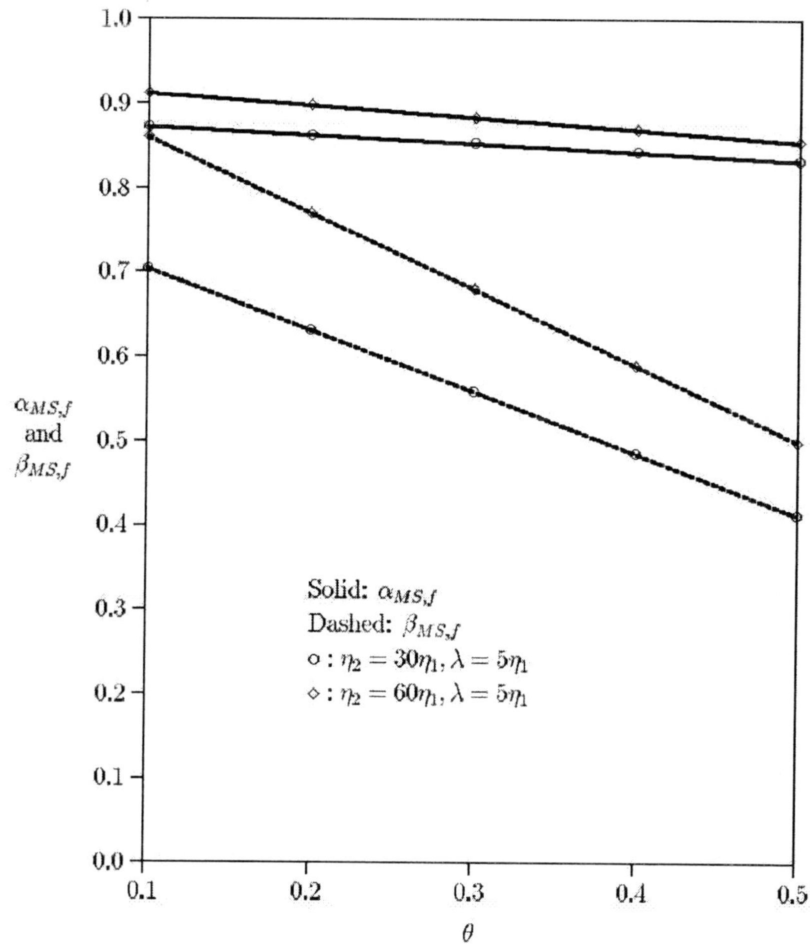

Figure 3: The Effect of η_2 (MS-DR with Fixed Registration Delay Period and IPP LA residence Times; $t_d = 1/(4\lambda)$, $\lambda = 5\eta_1$)

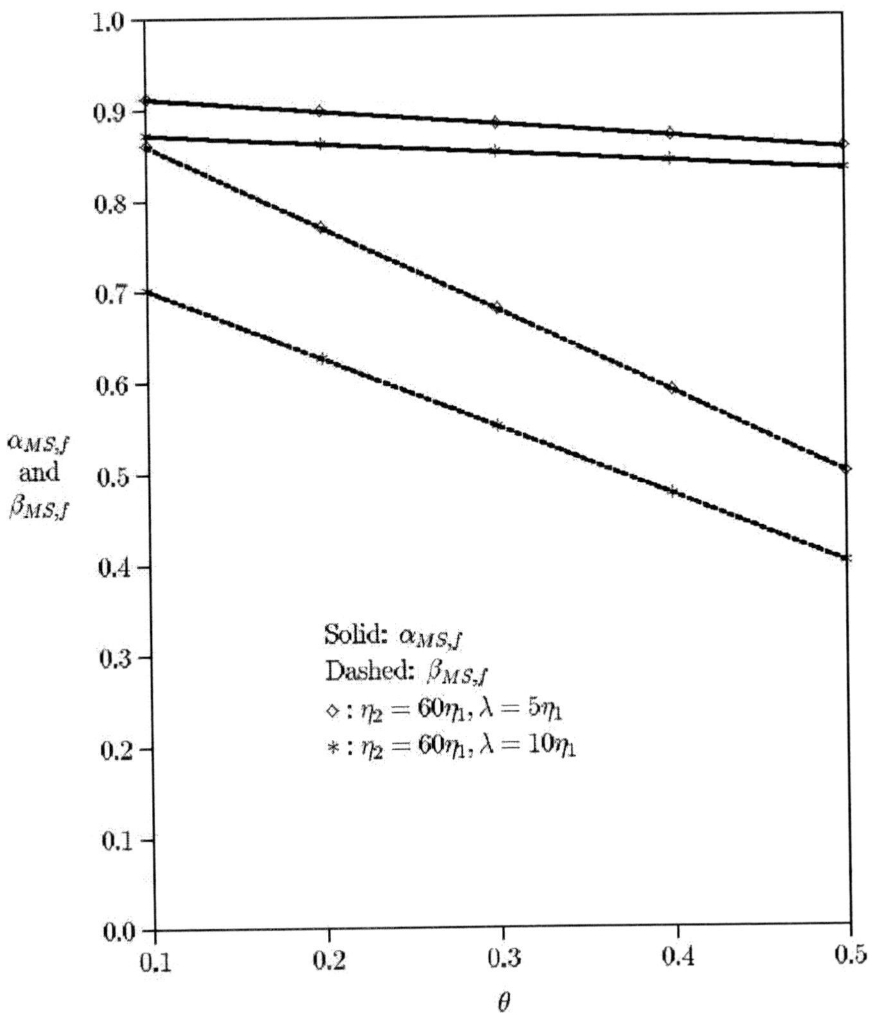

Figure 4: The Effect of λ (MS-DR with Fixed Registration Delay Period and IPP LA residence Times; $t_d = 1/(4\lambda)$, $\eta_2 = 60\eta_1$)

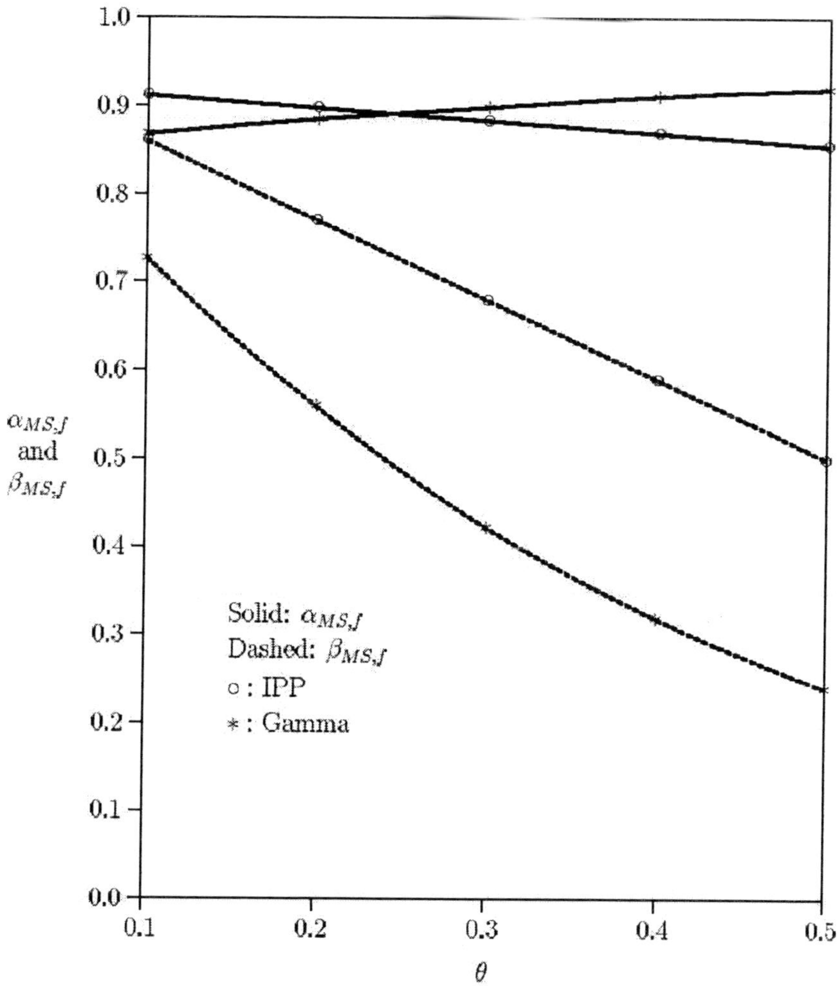

Figure 5: The Effect of the LA Residence Time Distributions (MS-DR with Fixed Registration Delay Period, $\eta_2 = 60\eta_1$, $\lambda = 5\eta_1$; $t_d = 1/(4\lambda)$)

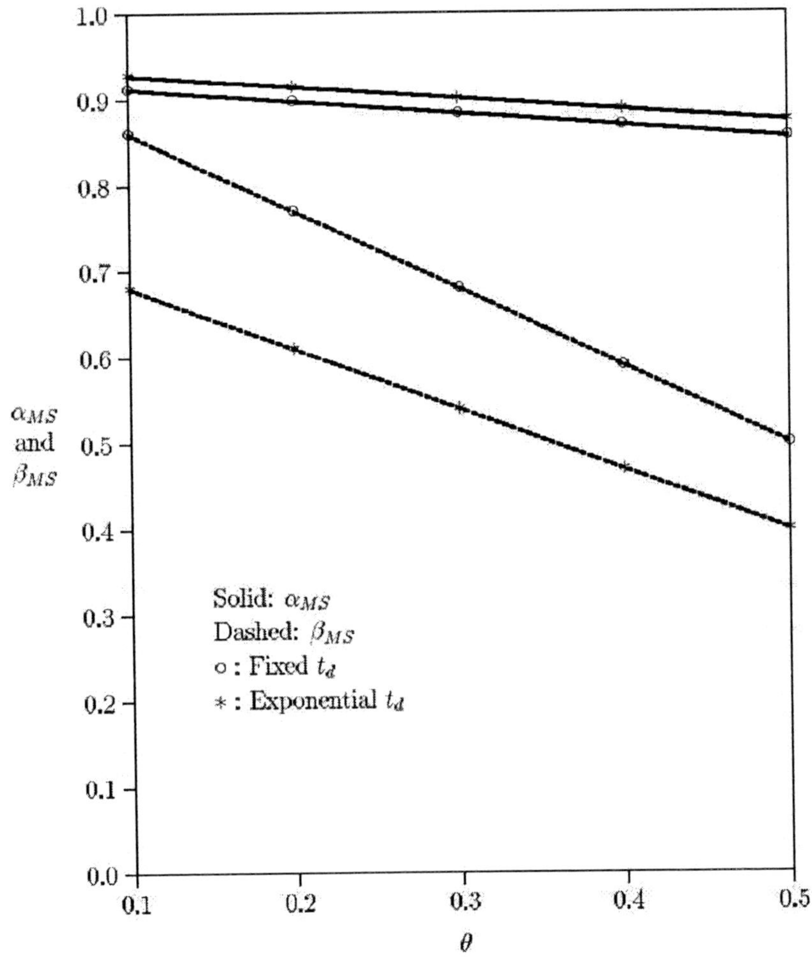

Figure 6: The Effect of the td Distributions (MS-DR with IPP LA Residence Times, $\eta_2 = 60\eta_1, \lambda = 5\eta_1$)

3 VLR-CONTROLLED DELAY REGISTRATION

This section models VLR-controlled delay registration (VLR-DR). To simplify our discussion, we assume that every VLR covers exactly one LA. Let τ_j be the time interval between the arrivals of the jth MS and the j + 1st MS. Then for the (N - n)th MS ($0 \le n \le N$), the registration delay period is

$$t_d = \sum_{j=0}^{n-1} \tau_{N-n+j}$$

Consider the example in Fig. 7. In this figure, N = 5. For MS_2, n=3 and its registration delay period is

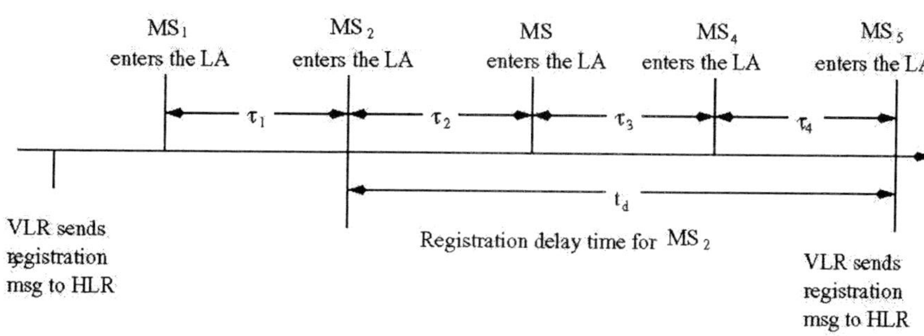

Figure 7: Registration Delay Period for VLR-DR

$$t_d = \sum_{j=0}^{2} \tau_{2+j} = \tau_2 + \tau_3 + \tau_4$$

Suppose that the arrivals of MSs to an LA can be represented by a Poisson process with rate ω (i.e., τ_j are exponentially distributed with mean 1/ω). Then for the (N - n)th MS, its registration delay period t_d has an Erlang density function

$$f_d(t_d) = \left[\frac{(\omega t_d)^n}{(n-1)!}\right] e^{-\omega t_d}$$

It is apparent that the portion β_{VLR} of the saved messages is

$$\beta_{VLR} = \frac{N-1}{N} \tag{17}$$

As illustrated in Fig. 1, the no-lost-call probability α_{VLR} for VLR-DR is expressed as

$$\alpha_{VLR} = \Pr[t_d \le t_c, t_d \le t_m] + \Pr[t_m \le t_c, t_m \le t_d] \tag{18}$$

where $\Pr[t_d \le t_c,\ t_d \le t_m]$ and $\Pr[t_m \le t_c,\ t_d \le t_m]$ are derived as follows.

$$\begin{aligned}\Pr[t_d \le t_c, t_d \le t_m] &= \int_{t_m=0}^{\infty}\int_{t_d=0}^{t_m}\int_{t_c=t_d}^{\infty} \lambda e^{-\lambda t_c}\left[\frac{(\omega t_d)^n}{(n-1)!}\right] e^{-\omega t_d} f_m(t_m)dt_c dt_d dt_m \\ &= \left(\frac{\omega}{\omega+\lambda}\right)^n \left\{1 - \sum_{j=0}^{n-1}\frac{[-(\omega+\lambda)]^j}{j!}\left[\frac{d^j f_m^*(s)}{ds^j}\right]\bigg|_{s=\omega+\lambda}\right\}\end{aligned} \tag{19}$$

and

$$\begin{aligned}\Pr[t_m \le t_c, t_m \le t_d] &= \int_{t_m=0}^{\infty}\int_{t_d=t_m}^{\infty}\int_{t_c=t_m}^{\infty} \lambda e^{-\lambda t_c}\left[\frac{(\omega t_d)^n}{(n-1)!}\right] e^{-\omega t_d} f_m(t_m)dt_c dt_d dt_m \\ &= \sum_{j=0}^{n-1}\left[\frac{(-\omega)^j}{j!}\right]\left[\frac{d^j f_m^*(s)}{ds^j}\right]\bigg|_{s=\lambda+\omega}\end{aligned} \tag{20}$$

From (19) and (20)

$$\begin{aligned}
\alpha_{VLR} &= \left(\frac{\omega}{\omega+\lambda}\right)^n - \sum_{j=0}^{n-1}\left(\frac{\omega}{\omega+\lambda}\right)^n \left\{\frac{[-(\omega+\lambda)]^j}{j!}\right\}\left[\frac{d^j f_m^*(s)}{ds^j}\right]\Bigg|_{s=\omega+\lambda} \\
&+ \sum_{j=0}^{n-1}\left[\frac{(-\omega)^j}{j!}\right]\left[\frac{d^j f_m^*(s)}{ds^j}\right]\Bigg|_{s=\omega+\lambda} \\
&= \left(\frac{\omega}{\omega+\lambda}\right)^n + \sum_{j=0}^{n-1}\left[\frac{(-1)^j}{j!}\right]\left[\omega^j - \frac{\omega^n}{(\omega+\lambda)^{n-j}}\right]\left[\frac{d^j f_m^*(s)}{ds^j}\right]\Bigg|_{s=\omega+\lambda}
\end{aligned} \tag{21}$$

For the IPP LA residence time t_m with the Laplace Transform expressed in (12), we have

$$\frac{d^j f_m^*(s)}{ds^j} = (-1)^j (j!)\left[\frac{\theta\eta_1}{(\eta_1+s)^{j+1}} + \frac{(1-\theta)\eta_2}{(\eta_2+s)^{j+1}}\right]$$

and the no-lost-call probability $\alpha_{VLR,IPP}$ is expressed as

$$\begin{aligned}
\alpha_{VLR,IPP} &= \left(\frac{\omega}{\omega+\lambda}\right)^n + \sum_{j=0}^{n-1}\left[\omega^j - \left(\frac{\omega}{\omega+\lambda}\right)^n(\omega+\lambda)^j\right]\left[\frac{\theta\eta_1}{(\eta_1+\omega+\lambda)^{j+1}}\right] \\
&+ \sum_{j=0}^{n-1}\left[\omega^j - \left(\frac{\omega}{\omega+\lambda}\right)^n(\omega+\lambda)^j\right]\left[\frac{(1-\theta)\eta_2}{(\eta_2+\omega+\lambda)^{j+1}}\right] \\
&= \left(\frac{\omega}{\omega+\lambda}\right)^n + \left(\frac{\theta\eta_1}{\eta_1+\lambda}\right)\left[1-\left(\frac{\omega}{\eta_1+\omega+\lambda}\right)^n\right] - \left(\frac{\omega}{\omega+\lambda}\right)^n\theta\left[1-\left(\frac{\omega+\lambda}{\eta_1+\omega+\lambda}\right)^n\right] \\
&+ \left[\frac{(1-\theta)\eta_2}{\eta_2+\lambda}\right]\left[1-\left(\frac{\omega}{\eta_2+\omega+\lambda}\right)^n\right] - \left(\frac{\omega}{\omega+\lambda}\right)^n(1-\theta)\left[1-\left(\frac{\omega+\lambda}{\eta_2+\omega+\lambda}\right)^n\right]
\end{aligned} \tag{22}$$

If f_m is a Gamma density function then its Laplace transform is expressed in (7) and

$$\frac{d^j f_m^*(s)}{ds^j} = \eta^\gamma(-1)^j\left[\frac{(\gamma+j-1)!}{(\gamma-1)!}\right](\eta+s)^{-(\gamma+j)}$$

From (21), the no-lost-call probability $\alpha_{VLR,G}$ is expressed as

$$\alpha_{VLR,G} = \left(\frac{\omega}{\omega+\lambda}\right)^n + \sum_{j=0}^{n-1}\binom{\gamma+j-1}{j}\left[\omega^j - \frac{\omega^n}{(\omega+\lambda)^{n-j}}\right]\left[\frac{\eta^\gamma}{(\eta+\omega+\lambda)^{\gamma+j}}\right] \tag{23}$$

Based on the analytic analysis, we use several numerical examples to study the performance of VLR-DR.

The effect of θ: Figures 8 and 9 (a) show that $\alpha_{VLR,IPP}$ is insensitive to θ. Thus, the number of short (or long) LA residence times in the IPP distribution do not affect the performance of VLR-DR.

The effects of ω: Fig. 8 (a) plots $\alpha_{VLR,IPP}$ with various ω values. If η_1 = 0.1 per hour, then $\omega = 200\eta_1$ means that the MSs arriving at an LA with the rate 20 per hour. The figure indicates that $\alpha_{VLR,IPP}$ is an increasing function of ω. For $\omega > 60$ per hour ($\omega > 600\eta_1$), $\alpha_{VLR,IPP}$ is only slightly improved as ω increases. In this figure, $\beta_{VLR,IPP} = (N - 1)/N = 50\%$.

The effect of N: Fig. 8 (b) shows that $\alpha_{VLR,IPP}$ is a decreasing function of N. The figure indicates that when N is small, increasing N results in slight degradation of $\alpha_{VLR,IPP}$ with large $\beta_{VLR,IPP}$ (= $\frac{N-1}{N}$) improvement. For example, by changing N from 2 to 3, $\alpha_{VLR,IPP}$ is degraded by 0.5%, and $\beta_{VLR,IPP}$ is improved by 33.3%. This phenomenon disappears when N is large. For example, by changing N from 4 to 5, $\alpha_{VLR,IPP}$ is degraded by 0.47%, and $\beta_{VLR,IPP}$ is only improved by 6.67%.

The effects of η_2 and λ: Fig. 9 (a) shows the effects of η_2 and λ. The results are similar to that for MS-DR, and will not be elaborated here.

The effect of the LA residence time distributions: Like MS-DR, Fig. 9 (b) indicates that the α_{VLR} performance for the Gamma LA residence time distribution is better than that for the IPP LA residence time distribution. Unlike MS-DR, $\beta_{VLR,IPP} = \beta_{VLR,G}$ for all N values.

To conclude, to achieve reasonable registration traffic reduction under the restriction of high no-lost-call probability, a small N (e.g., N = 2) should be selected.

4 Conclusions

We described two delay registration approaches to reduce the number of registration messages generated by PCS roaming management. In MS-controlled delay registration (MS-DR), after an MS enters an LA, no registration is issued by the MS until the registration delay period t_d expires.

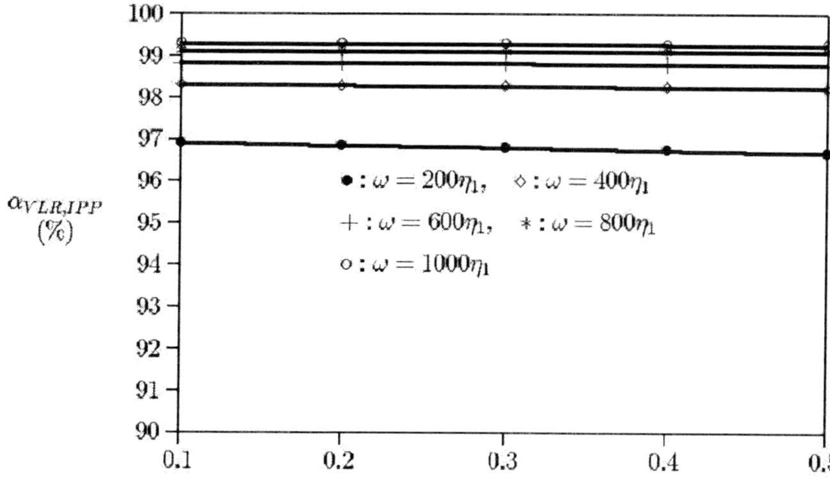

(a) The Effect of ω ($N = 2$)

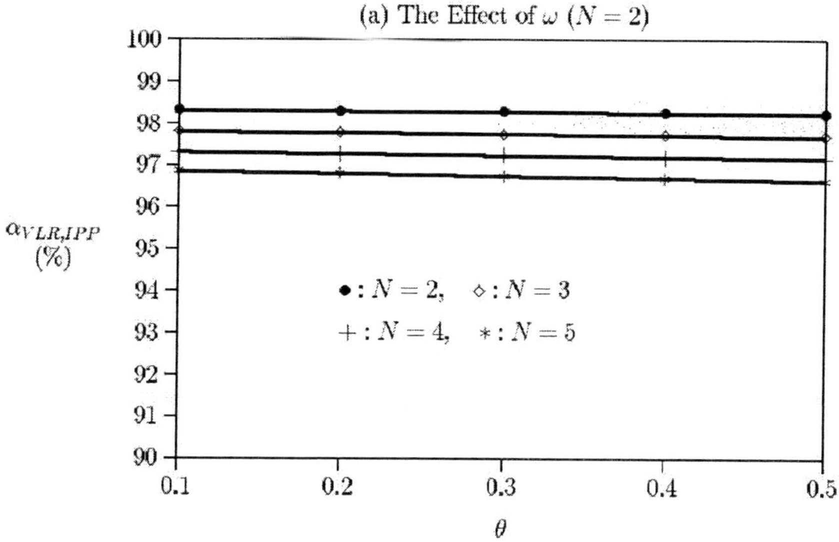

(b) The Effect of N ($\omega = 400\eta_1$)

Figure 8: The Performance of VLR-DR with IPP LA Residence Times
$(\eta 2 = \eta_2 = \eta_1,\ \lambda = 5\eta_1)$

If the MS moves out of the LA before t_d expires, then a registration message is saved. In MS-DR, call deliveries during the delay registration period are lost. Our study indicated that by selecting a small t_d, registration traffic can be effectively reduced at the cost of slightly decreasing the no-lost-call probability. Only minor modifications to MS are required to implement MS-DR. In modern MS design, the modifications can be done by remotely downloading a small piece of code from the mobile network to the MS through the radio link. The user will not notice the modification. No modifications to VLR/HLR are required. Also, MSs with MS-DR can co-exist with the traditional MSs (that do not implement MS-DR) without affecting their operations.

In VLR-controlled delay registration (VLR-DR), the VLR collects N registration messages (issued from the MSs) before it sends a registration notification message to the HLR. Like MS-DR, call deliveries to an MS during its delay registration period are lost in VLR-DR. Our study indicated that by selecting a small N value (e.g., $N = 2$), VLR-DR can effectively reduce the registration traffic with slight degradation of the no-lost-call probability. VLR-DR can be implemented by adding the capability of processing multiple component SS7 messages in VLR/HLR (this feature already exists in SS7 TCAP). No change to the mobility management protocols (such as IS-41 and GSM MAP) is required. No modification to MS is required. Traditional VLRs (that do not support VLR-DR) can co-exist with the HLR/VLRs that exercise VLR-DR.

As a final remark, we point out that the exercise of delay registration can be controlled by the network. When the network traffic is not heavy, the network may disable delay registration. On the other hand, the network may enable the delay registration mechanism during the traffic hours.

ACKNOWLEDGEMENT

Lin's work was sponsored in part by MOE Program of Excellence Research under contract 89-E-FA04-4, CCL/ITRI, Ericsson, InterVideo, FarEastone, National Science Council under contract NSC 89-2213-E- 009-203, the Lee and MTI Center for Networking Research, NCTU.

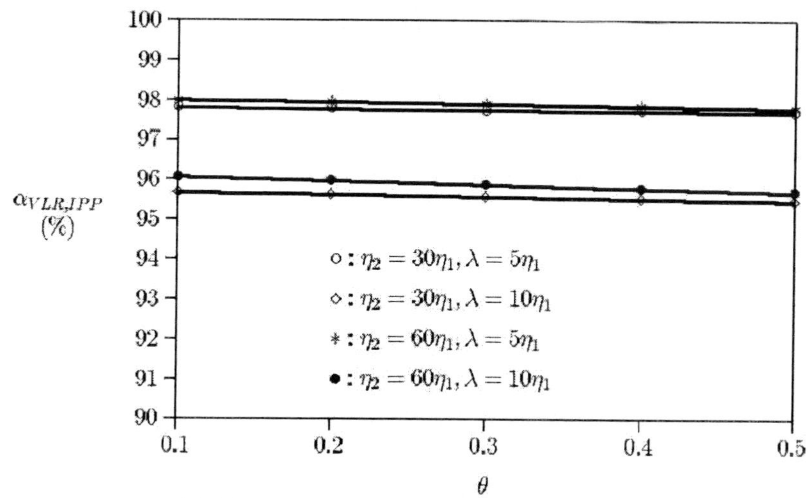

(a) The Effect of η_2 and λ (IPP LA Residence Times, $N = 2$)

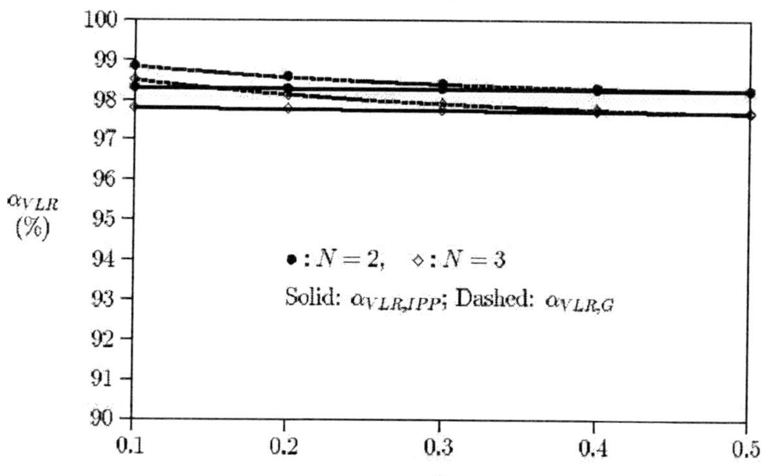

(b) The Effect of LA Residence Time Distributions ($\eta_2 = 30\eta_1, \lambda = 5\eta_1$)

Figure 9: The Performance of VLR-DR ($\omega = 400\eta_1$)

REFERENCES

[1] Akyildiz, I.F., Ho, S.M., And Lin, Y.-B. Movement-Based Location Update And Selective Paging for PCS Network. *IEEE/ACM Transactions on Networking*, 4(4):629–638, August 1996.

[2] ANSI. American National Standard for Telecommunications - Signaling System Number 7: Transaction Capabilities Application Part (TCAP), Issue 2 Rev. 2. Technical Report ANSI T1.114, ANSI, 1992.

[3] EIA/TIA. Cellular Intersystem Operations (Rev. C). Technical Report IS-41, EIA/TIA, 1995.

[4] ETSI/TC. Mobile Application Part (MAP) Specification, Version 4.8.0. Technical Report Recommendation GSM 09.02, ETSI, 1994.

[5] Grillo, D., Skoog, R., Chia, S., and Leung, K.K. Teletraffic Engineering for Mobile Personal Communications in ITU-T Work: The Need to Mth Practice and Theory. *IEEE Personal Commun. Mag*, 5(6):59–65, 1998.

[6] Jagerman, D.L. An Inversion Technique for the Laplace Transform. *Bell System Technical Journal*, 61(8):1995–2002, 1982.

[7] Johnson, N.L. And Kotz, S. *Distributions in Statistics: Continuous Multivariate Distributions*. John Wiley & Sons, 1972.

[8] Kuczura, A. The Interrupted Poisson Process As An Overflow Process. *Bell System Technical Journal*, 52, 1973.

[9] Lin, Y.-B., and Chlamtac, I. *Mobile and Wireless Network Architectures*. John Wiley & Sons, 2000.

[10] Lin, Y.-B. And DeVries S.K. PCS Network Signaling Using SS7. *IEEE Personal Commun. Mag.*, pages 44–55, June 1995.

[11] Lin, Y.-B., and Lin, P. Performance Modeling of Location Tracking Systems. *ACMMobile Computing and Communications*, 2(3):24–27, 1998.

Chapter 3

MOBILE TWO-TIER GATEWAYS FOR PCS SYSTEMS

Long-Sheng Li and Ming-Feng Chang
Department of Computer Science and Information Engineering
National Chiao Tung University
1001 Ta Hsuen Road, Hsinchu, Taiwan 30050,ROC
TAIWAN, R.O.C.
lsli@csie.nctu.edu.tw, mfchang@csie.nctu.edu.tw

Abstract: There are two major technologies in the personal communication service (PCS). High-tier PCS systems cover large continuous areas and support users moving at high speed. Low-tier PCS systems cover small zonal areas and only support users moving at low speed. In this paper, mobile two-tier gateways (MTTGs) are proposed to enable low-tier mobile users to use high-tier service so that low-tier users moving at high speed can also be served. The MTTGs can be installed on public transportation vehicles where low-tier service is unavailable due to high moving speed. The roaming management protocols for MTTGs and low-tier mobile users have been developed. The low-tier HLR database has been modified to support efficient registration and location tracking. Computer simulations were used to evaluate the performance of MTTGs. The simulation results indicate that low-tier mobile users using MTTGs experience higher blocking probability and forced termination probability than high-tier mobile users. The results also show that the two-tier interconnected system using MTTG extends the service area for low-tier users at the cost of degrading the quality of service for high-tier users.

Key-Words: - PCS, high-tier, low-Tier, MTTGs, blocking probability, forced termination probability.

1 INTRODUCTION

Personal Communication Service (PCS) systems can be divided into high-tier (cellular) systems and low-tier (cordless) systems depending on the technologies used. The base stations (BSs) of the high-tier (HT) PCS systems, such as AMPS (Advanced Mobile Phone Service) and GSM (Global System for Mobile Communication) [5., 7., 9.], use high radio transmitting power with antennas mounted on tall towers to cover a large area and support user speed up to 250 km/hr. On the other hand, the BSs of the low-tier (LT) PCS systems or cordless systems, such as DECT (Digital European Cordless Telecommunications), PACS (Personal Access Communications System), and CT2 (Cordless Telephone Second Generation) [5., 7., 9.], use low radio transmitting power to cover small and discontinuous areas. An HT PCS is cost-effective for users roaming at high speed because the coverage area is large and continuous. An LT PCS is cost-effective for high density environments such as metropolitan area and urban residential area where more communication channels are required.

The integration of two-tier PCS systems has been proposed recently [3., 11.]. The motivation to integrate HT and LT PCS systems is to provide the advantages of both systems. Depending on the network and the radio technologies, Lin [8.] classified multi-tier PCS systems into three groups: SRSN (similar radio technology, same network technology), DRSN (different radio technology, same network technology) and DRDN (different radio technology, different network technology). Examples of system integration include AMPS-PACS and GSM-PACS developed in Bellcore, and GSM-DECT in Europe [8.]. The two-tier PCS system can be constructed as low-tier microcell clusters overlaid with high-tier macrocells [1., 6., 14.]. Slow-moving mobiles are served by the microcells of the low-tier system, and fast-moving mobiles by the macrocells of the high-tier system [4., 10., 13., 14.]. The two-tier PCS still has some problems to overcome. For example, only dual-mode handsets can enjoy the benefit of the two-tier service. In addition, cell selection and call handoff between two tier systems is not easy to implement.

In Taiwan, the Directorate General of Telecommunications (DGT) has granted two low-tier wireless communication licenses in 1999. One company plans to adopt PHS (Personal Handy-phone System) [5., 7., 9.], and the other company plans to adopt PACS. In the near future, it is unlikely that an integrated two-tier system can be constructed. Moreover, dual-mode handsets capable of LT and HT communication are still rare. To support LT users moving at high speed and to extend the coverage of LT PCS system, we propose an alternative approach where low-tier handsets on public transportation vehicles can still be served through a two-tier gateway (TTG). A TTG is a communication gateway between an LT handset and an HT PCS network, so that LT handsets can use the HT service through a TTG when the LT service is unavailable. A TTG can be installed in public vehicles where the LT PCS is usually unavailable because the vehicle moves at high speed. As a result, the gateways are mobile as the vehicles move. The TTGs on moving vehicles will be referred to as mobile two-tier gateways (MTTGs). Unlike the two-tier PCS where dual-mode handsets are required to access two-tier service, an LT handset can access both LT and HT system services in our proposed system. The MTTGs also extend the coverage of the low-tier system, since the coverage of the high-tier system is large and continuous.

In this paper, we discuss the system architecture of PCS systems using MTTGs, propose a roaming management solution for the system, and evaluate the system performance using computer simulations. The paper is organized as follows. Section 2 describes the system architecture. Section 3 presents the roaming management of MTTGs. Section 4 discusses the performance issues of the system. Section 5 concludes the paper.

2 Mobile Two-Tier Gateways

Fig.1 illustrates a PCS architecture. A PCS system consists of a group of base stations (BSs) interconnected by a wireline network. The BS serves the calls to and from the mobile users in its service area, called cells, via radio links. One or more cells are grouped to form a registration area. The BSs in a registration area are connected to a mobile switching center (MSC) providing the call processing for all mobile users within the service area. To support user mobility, a two-level hierarchical location database is employed. The database consists of visitor location registers (VLRs) and a home location register (HLR). One or more MSCs are connected to a VLR. For each mobile user entering a registration area, the VLR keeps a record containing information on its location and service data. In addition, each mobile user has a corresponding HLR. For each mobile user, the HLR maintains a pointer pointing to the VLR where the mobile user visits.

Fig.2 shows the location information of a mobile user in the HLR and VLR. Note that every handset is assigned a Mobile Identification Number (MIN) in the location database. When an user subscribes to the PCS service, a record of user information is created in the HLR.

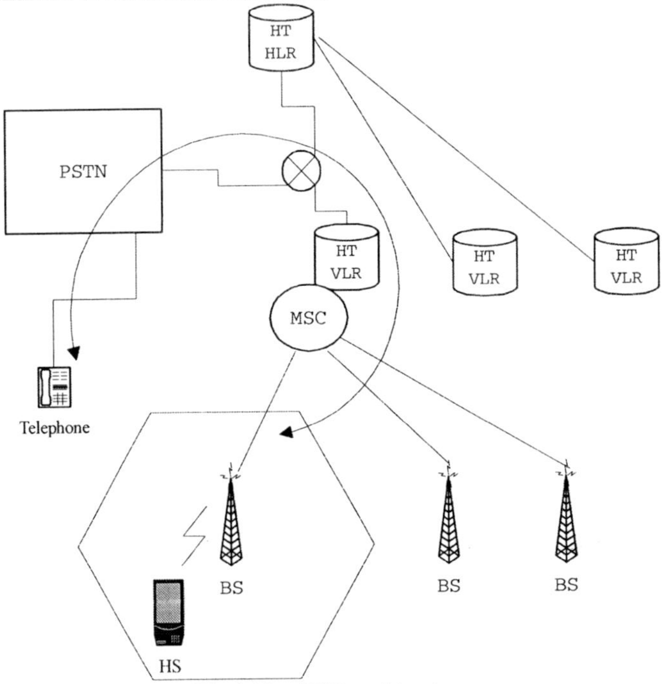

Fig.1 The PCS architecture.

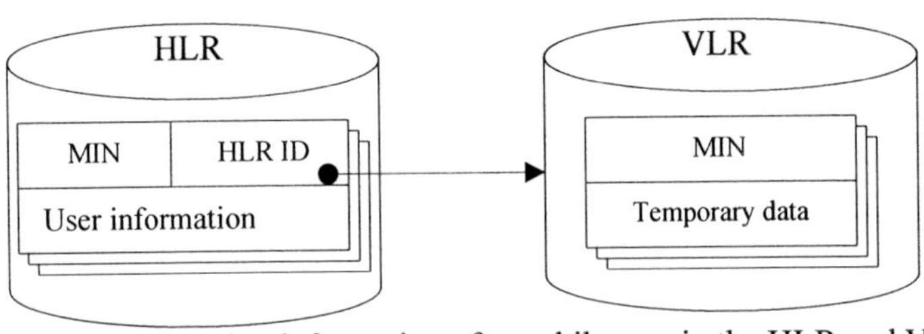

Fig.2 The location information of a mobile user in the HLR and VLR.

Consider the integration of a two-tier PCS system utilizing MTTGs. The HT and LT PCS systems can be interconnected by the public switched telephone network (PSTN). Fig.3 shows the interconnected network. An MTTG is the communication gateway between LT mobile users and the BSs of the HT system. An MSC functions like a switching node of the PSTN. Normally an LT mobile user communicates with a wireline phone through an LT BS, an LT MSC and the PSTN. Using the HT resource, the LT mobile user MIN2, shown in Fig.3, can communicate with a wireline phone through an MTTG, an HT BS, an HT MSC and the PSTN.

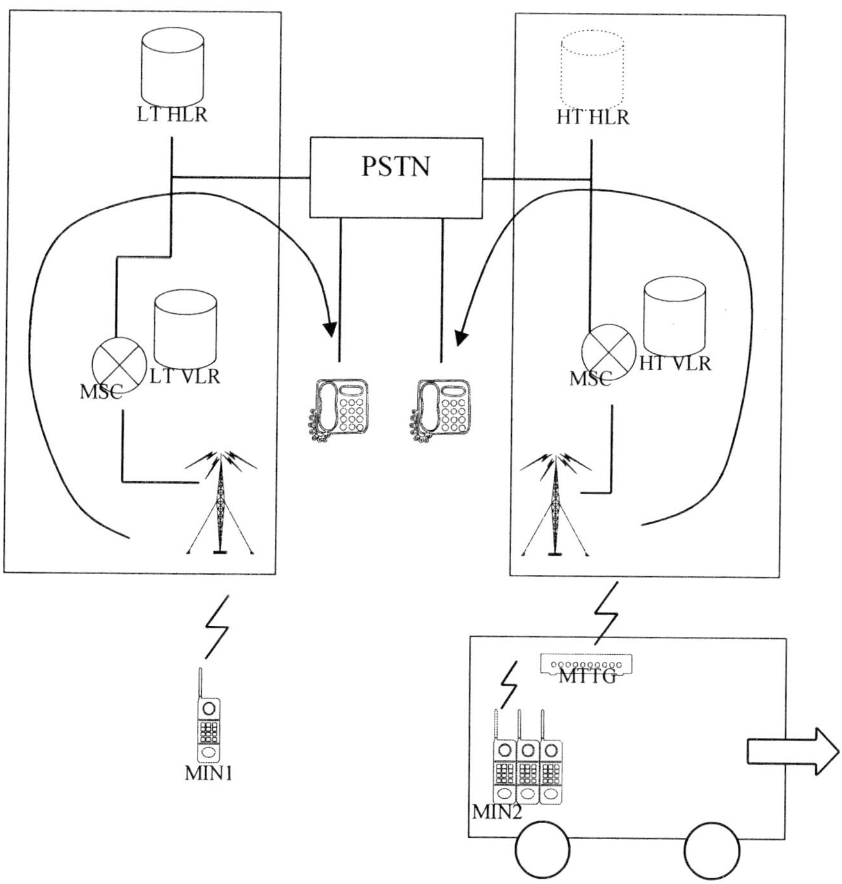

Fig.3 A two-tier PCS architecture using MTTGs.

An LT mobile user can initiate or receive a call through the LT or HT PCS systems. When an LT mobile user moves into a new registration area of the LT system, it registers to the LT system. The registration procedure remains unchanged; it is that of the LT system. On the other hand, when an LT mobile user roams into the service area of an MTTG, it registers to the MTTG. The MTTG registers the location of the mobile user to the LT HLR. This is a new registration procedure and will be described in next section.

Since an LT mobile user may communicate through an MTTG, the location of an MTTG should be tracked so that phone calls can be connected to the LT mobile users served by the MTTG. The LT HLR and the HT VLRs are used to track the location of an MTTG. The location information of an MTTG is stored in LT HLR and HT VLRs as shown in Fig.4. When an LT

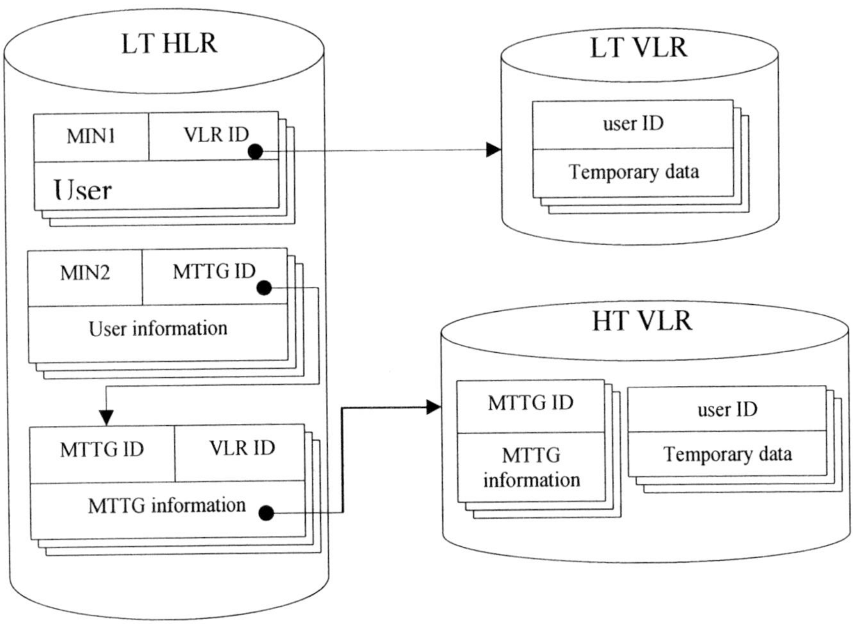

Fig.4 The location information of MTTGs and LT mobile users.

mobile user registers to an LT VLR, the mobile user location record points to the corresponding LT VLR; mobile user MIN1 in Fig.4 is an example. On the other hand, when an LT mobile user registers to an MTTG, the mobile user location record points to the MTTG; the MTTG location record in turns points to the corresponding HT VLR. For example, mobile user MIN2 in Fig.4 registers to an MTTG. Note that the location record of MIN2 in LT HLR is an indirect pointer.

3 ROAMING MANAGEMENT OF MTTG

In addition to modifying the location database, the roaming management of the PCS system using MTTGs requires three additional operations: LT mobile user registration through an MTTG, MTTG registration, and location tracking for an LT mobile user through an MTTG. The LT mobile user registration through an MTTG is a process where the LT mobile user informs the PCS system of its current location through an MTTG. The MTTG registration updates the location of the MTTG in the LT HLR. The location tracking for an LT mobile user may require the location information of an MTTG. Note that our proposed system does not support call handoff between the LT and HT systems. As a result, an LT user call is forced to terminate if the user moves from an LT registration area to an MTTG service area and vice versa.

3.1 LT Mobile User Registration through an MTTG

When an LT mobile user roams to an MTTG service area, the PCS system should modify the location record of the mobile user. As shown in Fig.5, mobile user **MIN1** moves from an LT registration area to an MTTG service area. Mobile user **MIN1** informs the HT VLR and LT HLR of its location. For example, if the HT system is GSM, this can be done using the Reserved element in Application Service Element (ASE) of the Signaling Connection Control Part (SCCP) of GSM Mobile Application Part (MAP) to carry the MTTG identity [9.]. The network message flow and protocol are shown in Fig. 6.

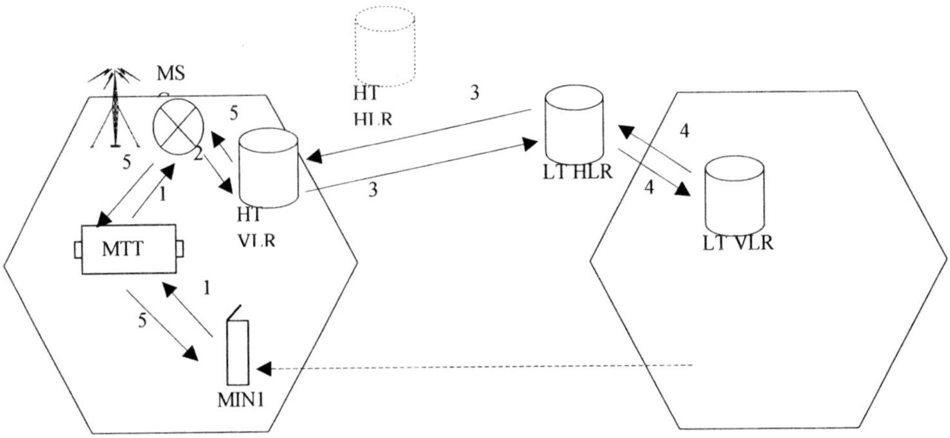

Fig.5 The procedure of an LT mobile user registration through an MTTG.

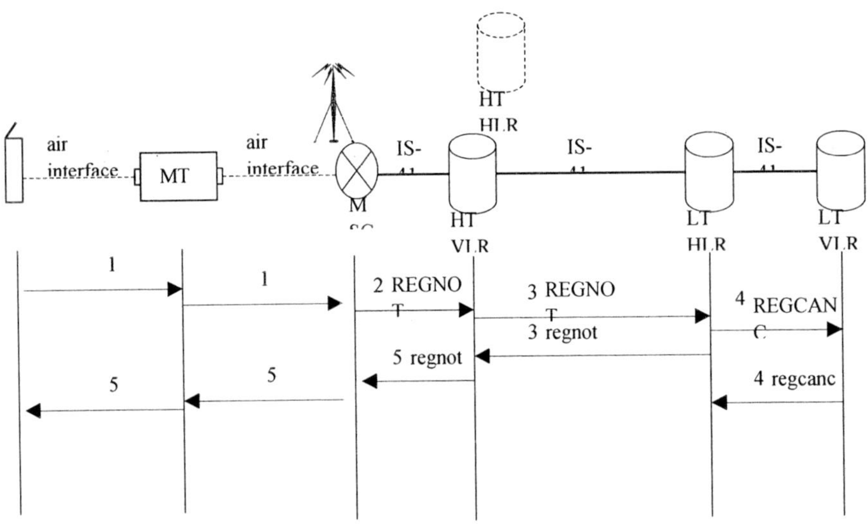

Fig.6 The message flow of an LT mobile user registration through an MTTG.

1. Mobile user **MIN1** registers to the MSC through the visited MTTG.
2. **REGNOT:** The corresponding MSC launches a registration request (REGNOT) to the HT VLR through Signal Transfer Points (STPs). Note

that STPs are switches that relay SS7 messages between telephone switches and databases [9.].

3. **REGNOT, regnot:** The HT VLR informs the LT HLR of the LT mobile user's current location (sends an IS-41 registration notification message, REGNOT) and then the LT HLR sends an acknowledgement (regnot) to the HT VLR. The HT VLR may not know the actual address of LT HLR. Instead, HT VLR sends the message with the MIN of the mobile user to an STP which can translate the MIN into the HLR address by a table lookup technique called Global Title Translation (GTT).
4. **REGCANC, regcanc:** The LT HLR sends a deregistration message (REGCANC) to the LT VLR where the mobile user previously visited to cancel the obsolete registration record. In addition, the LT VLR sends an acknowledgement (regcanc) back to the LT HLR.
5. **regnot:** The HT VLR sends a **regnot** acknowledgement to the MSC. This informs the MSC of the completion of the location registration in the HT VLR. The MSC then informs the MTTG of the registration result, and the MTTG informs the LT mobile user.

3.2 MTTG Registration

When an MTTG moves from an HT registration area to another, an MTTG registration procedure is activated and the PCS system updates the location record of the MTTG. The steps are shown in Fig.7.

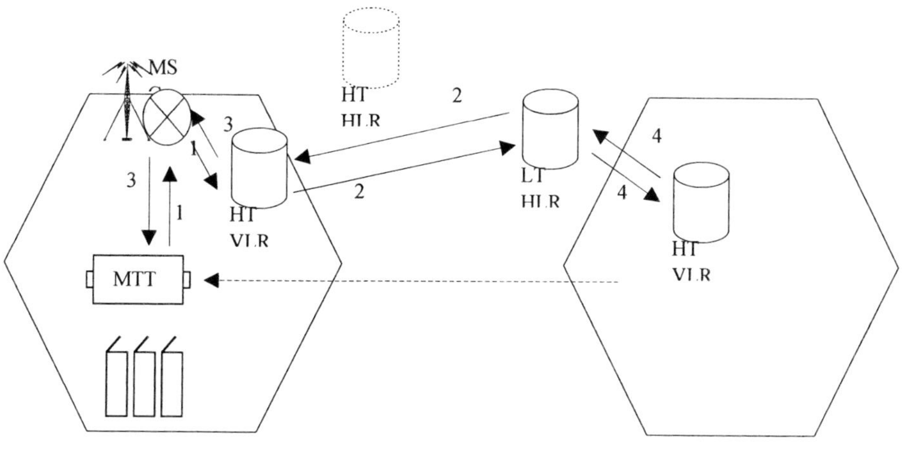

Fig.7 The procedure of MTTG registration.

1. **MREGNOT:** In order to follow the IS-41 protocol, the corresponding MSC launches a registration query to the HT VLR through STPs.
2. **MREGNOT, mregnot:** The new HT VLR informs the LT HLR of the MTTG's current location, and the LT HLR sends an acknowledgement to the new HT VLR. When the LT HLR receives the registration message, it modifies the MTTG location information.
3. **mregnot:** The new HT VLR informs the MTTG of the successful registration.
4. **MREGCANC, mregcanc:** After step 2, the LT HLR sends a deregistration message (MREGCANC) to cancel the location record in the old HT VLR. When the old HT VLR receives the deregistration message, it deletes the MTTG location record.

Note that in this registration, the MTTGs register to the new HT VLR and LT HLR, but the mobile users on the MTTGs do not register their location information to the new HT VLR. As a result, the network load of the registration is reduced. However, when an LT mobile user originates a call from an MTTG, the HT VLR should obtain the mobile user information from the LT HLR because the HT VLR does not maintain any record of LT users.

3.3 Location Tracking for LT Mobile Users

For a call to an LT mobile user, the PCS system must locate the mobile user. A mobile user can be located in an LT registration area or in an MTTG service area. If an LT mobile user visits an LT registration area, the location tracking process is the same as that of the original LT system. If it visits an MTTG service area, the location tracking process, shown in Fig.8, can be described as the following steps.

1. The call is forwarded to an originating switch in PSTN. Through the GTT mechanism in the STPs, the originating switch sends a message to the LT HLR to query the location of the LT mobile user. The LT HLR determines that the LT mobile user is in an MTTG service area. The LT HLR maintains the MTTG location record pointing to the HT VLR serving the MTTG. The LT HLR queries the HT VLR for the MTTG current location.

2. The HT VLR returns a routable address to the originating switch through the LT HLR.
3. Based on the routable address, a trunk (voice circuit) is then set up from the originating switch to the visited MSC.

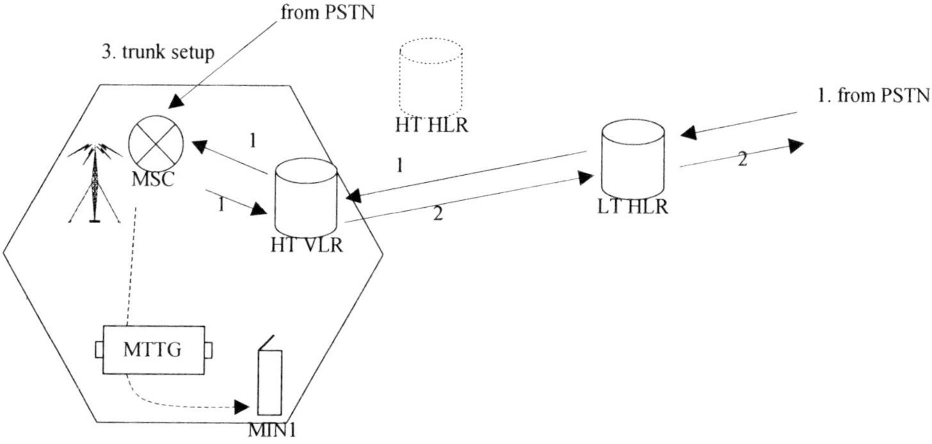

Fig.8 The procedure of location tracking.

Before the trunk is set up, the handset should request an LT channel and the MTTG should request an HT channel to complete the call setup. Therefore, after the BS pages the handset via the MTTG, the handset sends a channel-request message to the MTTG. If there is no free LT channel, the MTTG sends an access-denied message to the handset and does not send any message to the BS. In this case, the BS does not receive the channel-request message within a timeout period and this call is rejected. Otherwise, when the MTTG has a free LT channel, the MTTG sends a channel-request message to the BS to request an HT channel. If there is no available HT channel, the BS sends an access-denied message to the MTTG and the MTTG sends an access-denied message to the handset. The call then is rejected. Otherwise, when the BS has a free HT channel, the BS sends an access-granted message to the MTTG and the MTTG sends an access-granted message to the handset. The channels are assigned and the trunk is set up.

3.4 Call Origination for LT Mobile Users

When the serving MSC receives a call origination request from an LT mobile user through an MTTG, the call initialization steps are shown in Fig.9. Since the HT VLR does not have any data of the caller, the serving MSC should request the user information from the LT HLR to determine whether the call is allowed.

1. The serving MSC informs its associated HT VLR and requests the user validation information from the LT HLR.
2. The LT HLR returns the requested information to the MSC and then the MSC determines whether the call is allowed according to the user validation information.
3. If the call is allowed, voice trunk will be set up for the call.

Before the trunk is set up, the handset should request an LT channel and an HT channel using the protocol without the paging procedure as described before.

4 Simulation and Performance Evaluation

An LT mobile user connected through an MTTG requires two radio channels: one LT channel and one HT channel. In addition, when an MTTG moves into a new HT cell, all the connected LT mobile users on the MTTG should be handed over to the new cell. The handoff may require more than one free channel in the cell; this handoff will be referred to as multi-channel handoff. We expect that the LT mobile users on an MTTG experience more forced terminations due to multi-channel handoff. In this section, we will study the service quality of the LT mobile users using MTTGs.

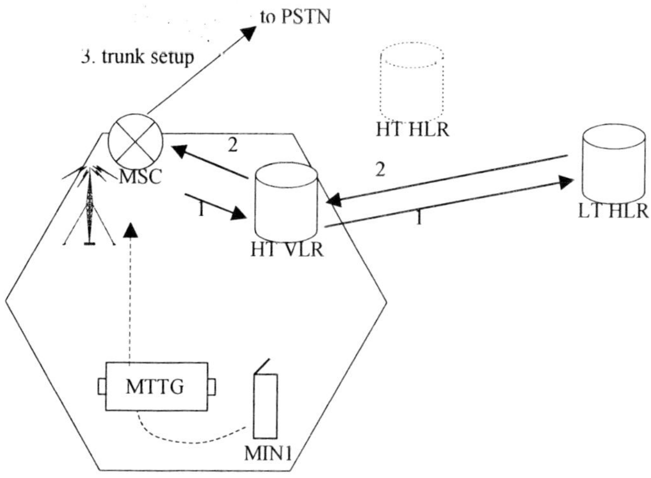

Fig.9 The procedure of call origination.

First, we study the LT channel assignment problem on an MTTG. Assume the call arrivals on an MTTG form a Poisson process. Let C_m be the channel capacity of an MTTG, λ_0 be the call arrival rate on the MTTG, the call holding time be an exponential distribution with mean $1/\mu_0$. This problem can be modeled as an M/M/C queue. Let π_i be the steady state probability that there are i connecting calls on the MTTG. In the steady state, the call arrival rate equals the call completion rate and the sum of all state probabilities equals 1 [2.]. From the equations, π_0 and π_i can be computed as

$$\pi_0 = \frac{C_m!}{C_m! + \sum_{i=1}^{C_m} \frac{C_m!}{i!} (\frac{\lambda_0}{\mu_0})^i} \tag{1}$$

$$\pi_i = \frac{1}{C_m!} \left(\frac{\lambda_0}{\mu_0} \right)^i \pi_0, \text{for } i = 1, \cdots, C_m \tag{2}$$

From equations (1) and (2), we can obtain the new call blocking probability ($= \pi_{C_m}$) and the channel utilization ($= \frac{\sum_{i=1}^{C_m} i\pi_i}{C_m}$).

To provide acceptable service quality, the call blocking probability should be less 2%. Fig.10 and Fig.11 illustrate the minimum requirement of channel capacity and channel utilization of an MTTG with the blocking probability less then 2% for various traffic load (λ_0/μ_0). The expected call holding time ($1/\mu_0$) is assumed to be 3 min. The experiment results show that at least 4 radio channels are required on the MTTG when the traffic load is equal to 1. In this case, the channel utilization on the MTTG is only 25%. The minimum number of channel requirement increases as the traffic load increases. In addition, the channel utilization improves as the traffic load increases. However, the channel utilization is less than 50% when the traffic load is less than 5. We would expected the channel utilization of the MTTGs installed on public transportation vehicles to be low because little traffic load would be generated from the limited number of LT users on the vehicle.

Second, we study the HT channel assignment problem. We consider an HT PCS system that serves HT mobile users and LT mobile users using MTTGs. Since no feasible queueing model can be used to analyze this problem, event-driven computer simulations were used. In the simulations, 5×5 cells are connected as a wrap-around mesh. For each cell, let the number of channels be C_C and the average number of MTTGs be n_m.

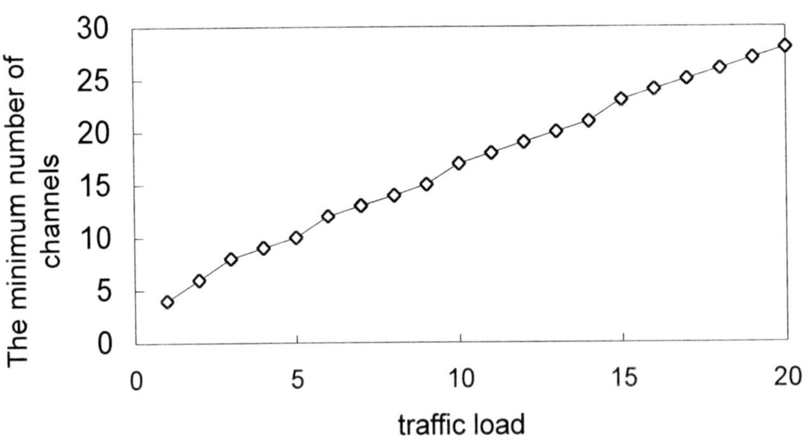

Fig.10 The minimum number of channels in an MTTG with the blocking probability less than 2%.

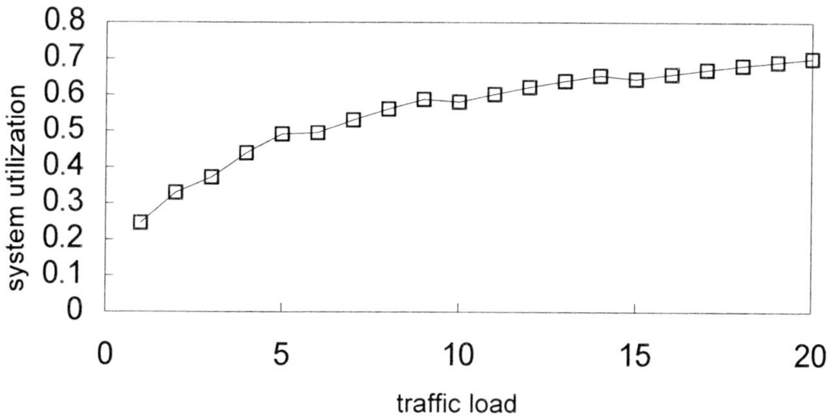

Fig.11 The channel utilization in the MTTG with the blocking probability less than 2%.

Fig.12 shows the flow chart of the simulation program. At the beginning of the simulations, three types of events are initialized and pushed into an event queue in a time-increasing order; the events include a call arrival event of an HT mobile user (HUarrival) for each cell, a handoff event for each MTTG (Mhandoff), and a call arrival event of each MTTG (MUarrival). The main iteration of the simulation program begins by pulling out the first event from the event queue. The events are processed based on the event type. Consider the HT mobile users, there are three types of events to be processed. If the pulled event is an HUarrival event, the HUarrival procedure will be performed. The procedure first generates a new HUarrival event and pushes the event into the event queue. Then, it generates a handoff time which is exponentially distributed and compares it with the call completion time. If the handoff time is less than the call completion time, the call will be handed over to another cell. In this case, the procedure generates an HUhandoff event (a call handoff event of an HT mobile user) and decides the next cell where the call will be handed over, otherwise, it generates an HUcompletion event (a call completion event of an HT mobile user). If the pulled event is the

HUhandoff event, the simulation program generates a handoff time and compares the time with the call completion time. Based on the comparison result, an HUhandoff event or an HUcompletion event will be generated and put into the event queue. If the pulled event is an HUcompletion event, adds a call completion into the statistic data.

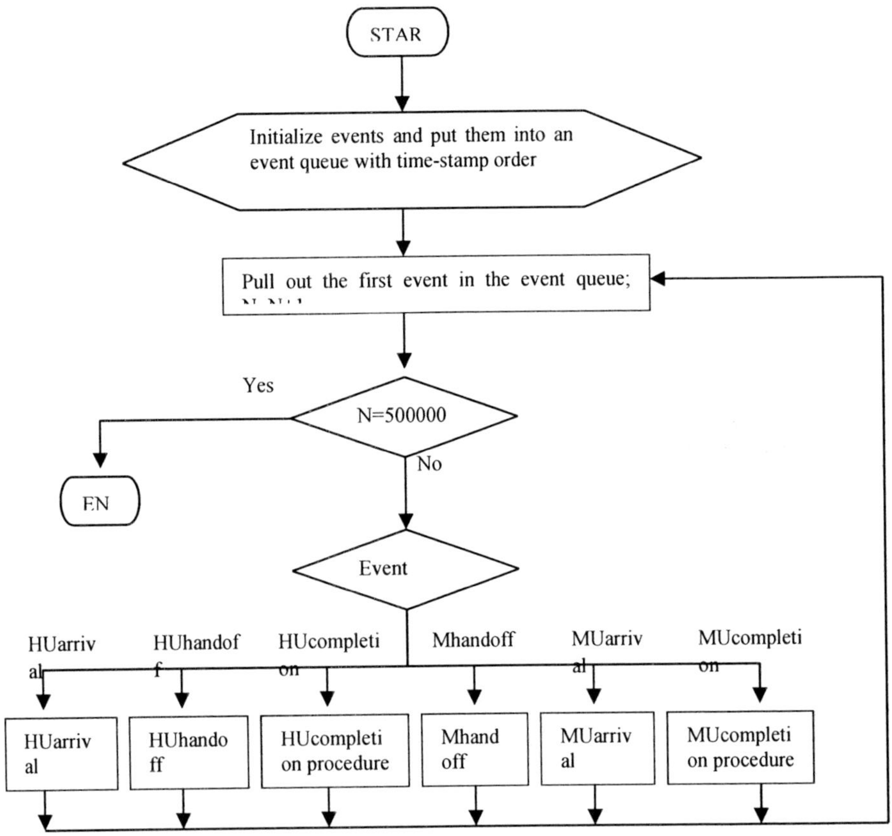

Fig.12 The simulation algorithm.

Consider the MTTG users, there are two types of events to be processed: the call arrival event (MUarrival) and the call completion event (MUcompletion). The MUarrival procedure is similar to the HUarrival procedure. It initializes another Muarrival and generates an MUcompletion event. The MUcompletion procedure is the same as the HUcompletion procedure. It adds an MTTG user call completion into the statistic data.

When an MTTG moves to another HT registration area, the users using the MTTG also move to the registration area. The event is denoted as an Mhandoff event. The Mhandoff procedure decides the next cell to which the MTTG will move and generates the next Mhandoff event. A counter N is used to count the number of events processed. The simulation iteration terminates if the counter N=5000000.

The simulation results described below were obtained by setting the initial parameters as follow. The call holding time of HT mobile users is exponentially distributed with mean $1/\mu_{hu}$ = 3 min., the residence time of HT mobile users in an HT cell is exponentially distributed with mean $1/\eta_{hu}$ = 15 min., the residence time of MTTGs in an HT cell is exponentially distributed with mean $1/\eta_m$ = 15 min., and the call holding time of MTTG users is exponentially distributed with mean $1/\mu_{mu}$ = 3 min. The HT mobile user traffic load is denoted as $\rho_{hu} = \frac{\lambda_{hu}}{\mu_{hu}}$, where λ_{hu} is the call arrival rate of HT mobile users. The user load on an MTTG is denoted as $\rho_{mu} = \frac{\lambda_{mu}}{\mu_{mu}}$, where λ_{mu} is the call arrival rate of MTTG users. Both the call arrival of HT mobile users and the call arrival of MTTG users form a Poisson process. The system load ρ is the sum of HT mobile user load and MTTG user load, $\rho = \rho_{hu} + \rho_{mu} \times n_m$, where n_m is the average number of MTTGs in a cell.

First, we study the effect of MTTG user load on the system performance. In the simulations, the net system traffic load including HT mobile users and MTTG users is 10. The expected traffic load in an MTTG is 2.5. To reduce the effects of the limited number of LT channels in the MTTGs on the call blocking probability of LT mobile users, let the channel capacity of an MTTG be infinite. The average number of MTTGs (n_m) per cell varies from 0 to 4 for various MTTG user loads. Thus, the average traffic load of the MTTG users is varied as 0%, 25%, 50%, 75%, and 100% of the net system traffic load. However, the net system load remains the same.

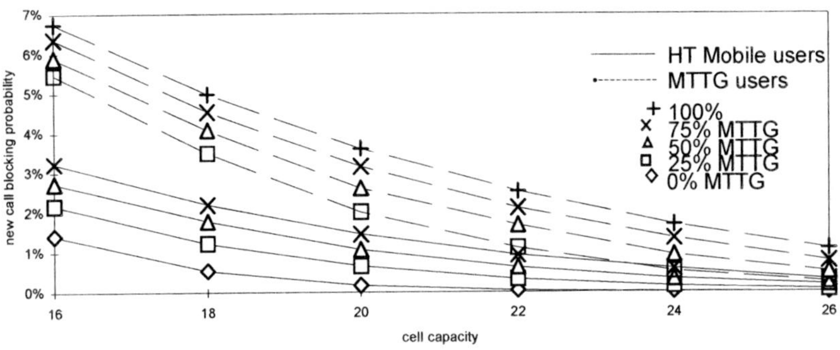

Fig.13 The new call blocking probability for various MTTG traffic.

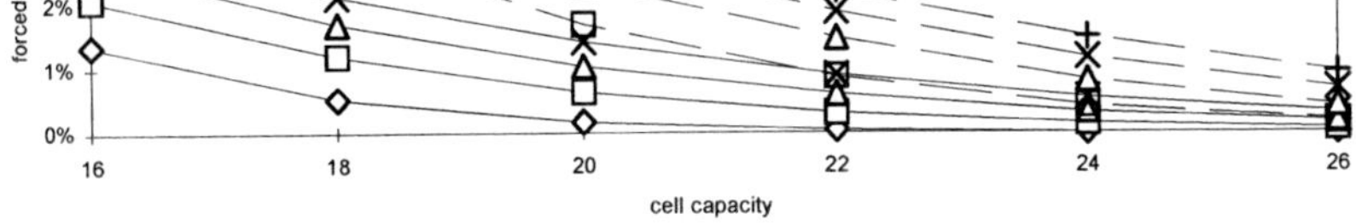

Fig. 14 The forced termination probability for various MTTG traffic.

The results in Fig.13 show that the blocking probability of the MTTG users is higher than that of the HT mobile users. Since an MTTG gathers a number of connected calls, a cell is more likely to run out of free channels if MTTGs are present. As a result, the LT users using MTTGs experience more calls blocked. For the same net system load, the results also show the blocking probability of HT mobile users increases as the percentage of MTTG users increases.

Fig. 14 shows the effect of multi-handoff. The forced termination probability of MTTG users is higher than that of the HT mobile users. This is because more than one free channel is required during multi-channel handoff. Thus, the MTTG users experience more forced terminations during handoff. As a result, the call incompletion probability (a probability that a call is either blocked or forced terminated) of MTTG users is higher than that of HT mobile users as shown in Fig.15. Note that as the portion of MTTG users increases, the forced termination probabilities of both HT users and MTTG users increase. All the simulation results indicate the system performance degrades due to the presence of the MTTG users.

CONCLUSION

We have proposed MTTGs which can be installed on transportation vehicles to enable LT PCS users to use HT PCS service. A two-tier interconnected system architecture using MTTGs has been described. In our design, the MTTGs are transparent to the LT users. To support the roaming management for LT PCS users using MTTG service, the LT HLR has been modified. Moreover, an efficient MTTG registration procedure and an LT user location tracking procedure have been developed. Computer simulations were used to evaluate the system performance. The simulation results show that the new call blocking probability, the forced termination probability, and the call incompletion probability of MTTG users are higher than those of HT mobile users. The simulation results also show that for the same net traffic load, the call blocking probability and the call incompletion probability of HT users increase as the percentage of MTTG users increases. In summary, the two-tier interconnected system using MTTG extends the service area for LT users at the cost of degrading the quality of service for HT users.

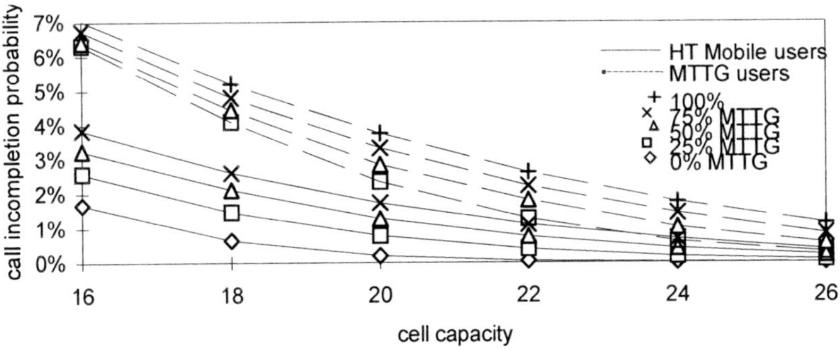

Fig.15 The call incompletion probability for various MTTG traffic.

REFERENCES

[1.] M. Benveniste, Cell Selection in Two-tier Microcellular/Macrocellular Systems, *IEEE Global Telecommunications Conference 1995,* Vol.1, 1995, pp.1532-1536.

[2.] R. B. Cooper, *Introduction to Queueing Theory,* 2nd ed. New York: Elsevier North Holland, 1981.

[3.] D. C. Cox, Wireless Personal Communications: What Is It? *IEEE Personal Communications Magazine,* April 1995, pp.20-35.

[4.] A. Grimaldi, S. Marano, C. Mastroianni and M. T. Scarpelli, A Macro-Microcellular System with Mobile User Speed Estimation, *The 8th IEEE International Symposium on Waves of the Year 2000, PIMRC'97,* Vol.2, 1997, pp.381-385.

[5.] A. D. Kucar, Mobile radio: An Overview, IEEE Communications Magazine, Nov. 1991, pp.72-85.

[6.] X. Lagrange and P. Godlewski, Performance of a Hierarchical Cellular Network with Mobility-Dependent Hand-Over Strategies, *IEEE 46th Mobile Technology for the Human Race, Vehicular Technology Conference 1996,* Vol.3, 1996, pp.1868-1872.

[7.] O. K. Li and X. Qiu, Personal Communication Systems (PCS), *Proceeding of the IEEE,* Vol.83, No.9, Sept. 1995, pp.1210-1243.

[8.] Yi-Bing Lin and I. Chlamtac, Heterogeneous Personal Communications Services: Integration of PCS Systems, *IEEE Communications Magazine,* Sept. 1996, pp.106-113.
[9.] Yi-Bing Lin, *Introduction to Mobile Network Management,* Baltzer Publisher, 1997.
[10.] S. Nanda, Teletraffic Models for Urban and Suburban microcells: Cell sizes and handoff rates, *IEEE Transactions of Vehicular Technology,* Vol.42, No.4, Nov. 1993, pp.673-682.
[11.] J. F. Rizzo, and N. Sollenberger, Multitier Wireless Access, *IEEE Personal Communications Magazine,* 1995, pp.18-31.
[12.] A. Samukic, UMTS Universal Mobile Telecommunications System: Development of Standards for the Third Generation, *IEEE Transactions on Vechnology,* Vol.47, No.4, Nov. 1998, pp.1099-1104.
[13.] C. W. Sung and W. S. Wong, User Speed Estimation and Dynamic Channel Allocation in Hierarchical Cellular System, *1994 IEEE 44th Vehicular Technology Conference,* Vol.1, 1994, pp.91-95.
[14.] K. L. Yeung and S. Nanda, Channel Management in Microcell/Macrocell Cellular Radio Systems, *IEEE Transactions of Vehicular Technology,* Vol.45, No.4, Nov. 1996, pp.601-612.

Chapter 4

MARKOV MODULATED POISSON PROCESS MODEL FOR HAND-OFF CALLS IN CELLULAR SYSTEMS

M. Guizani* and M. A. Farahani**
Computer Science Department
University of West Florida
**Electrical and Computer Engineering
University of Missouri - Columbia
UNITED STATE OF AMERICA
mguizani@uwf.edu and Mfarahani@kansascity.com

Abstract: The growth of telecommunication system is more than ever stimulating the need for mathematical modeling and evaluation techniques to predict the performance of these systems. These techniques have become an indispensable tool for assessment of the performance of a system, evaluation of design alternatives, resource dimensioning, and system configuration. Poisson distribution is a good representative for most of the telecommunication events. The hand-off calls in cellular system have bursty characteristic which means they are happening with different rates. Therefore, there is a need for a model that is able to capture this specific behavior. In this paper, Markov Modulated Poisson Process (MMPP) is used to model the hand-off calls. Then, an analytical model of a cellular system is solved and the performance metrics are calculated. Finally we compared our results with those obtained when considering only the Poisson model.

Key-Words: - Cellular system, Handoff calls, New calls, Poisson Distribution, Markov Modulated Poisson Process.

1 INTRODUCTION

1.1 Cellular System

The ability to communicate with people on the move has evolved remarkably since Guglielmo Marconi first demonstrated radio's ability to provide continuous contact with ships sailing the English Channel. That was in 1897, and since then new wireless communications methods and services have been enthusiastically adopted through out the world. Most people are familiar with a number of mobile radio communication systems used in every day life; the cellular telephones are one of them. The term mobile has historically been used to classify any radio terminal that could be moved during operation. In general, the collective group of users in a wireless system is called mobiles or users. The mobiles communicate to fixed base stations. The base stations are located at the center or on the edge of a coverage region and consist of radio channels as well as transmitter and receiver antennas mounted on a tower [4].

In the following sections, we will present the basic characteristics of the cellular telephone systems: cellular telephone system, properties of cellular geometry, function description of system operation, hand-off strategies, and prioritizing hand-off calls.

1.1.1 Cellular Telephone System

A cellular telephone system provides a wireless connection to the Public Switched Telephone Network (PSTN) for any user location within the radio range of the system. Cellular systems accommodate a large number of users over a large geographic area, within a limited frequency spectrum. Cellular radio systems provide high quality service that is often comparable to that of the land line telephone systems. High capacity is achieved by limiting the coverage area called a cell, so that the same radio channels may be reused by another base station located some distance away. A sophisticated switching technique, called a hand-off, enables a call to proceed uninterrupted when the user moves from one cell to another [4].

1.1.2 Properties of Cellular Geometry

The main purpose of defining cells in a mobile-telephone system is to delineate areas in which either specific channels or a specific cell site will be used at least preferentially, if not exclusively. A reasonable degree of geographical confinement of channel usage is necessary to prevent co-channel interference problems. Having defined a desired cellular pattern in concept, system planners achieve that pattern in the field through proper positioning of land transmitter sites, proper design of the azimuthal gain pattern of the site's antennas, and proper selection during every call of a suitable site to serve the call.

The irregular land transmitter spacing and amorphous cell shapes shown in fig.1 might be acceptable in a system where the initial system configuration, including the selection of transmitter sites and the assignment of channels to cells, could be frozen indefinitely. In practice, however, the absence of an orderly geometrical structure in a cellular pattern would make adaptation to traffic growth more cumbersome than necessary. Inefficient use of spectrum and uneconomical development of equipment would be likely outcomes. A great deal of improvisation and custom engineering of radio, transmission, switching, and control facilities would be required repeatedly in the course of system growth.

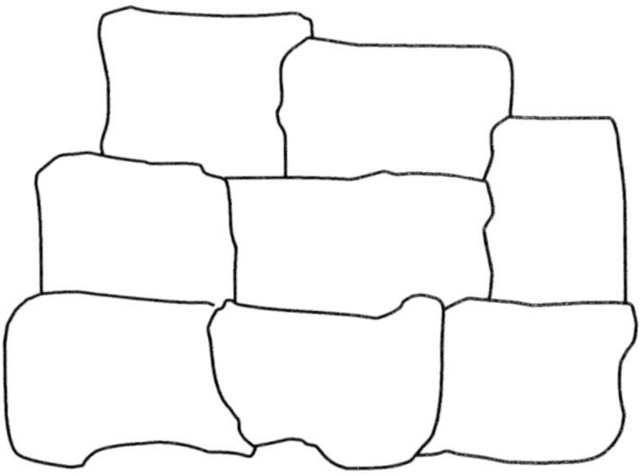

Fig.1: Cellular layout illustrating frequency reuse

Early in the evolution of the cellular concept, system designers recognized that visualizing all cells as having the same shape helps to systematize the design and layout of cellular systems. A cell was viewed as the coverage area of a particular land site. If, as with present-day mobile service, omnidirectional transmitting antennas were used, then each site's coverage area-bounded by contour of constant signal level-would be roughly circular. Although propagation considerations recommend the circle as a cell shape, the circle is impractical for design purposes, because an array of circular cells produces ambiguous areas, which are contained either in no cell or in multiple cells. On the other hand, any regular polygon approximates the shape of a circle and three types, the equilateral triangle, the square, and the regular hexagon can cover a plane with no gaps or overlaps as shown in fig.2. A cellular system could be designed with square or equilateral triangular cells, but, for economic reasons, Bell Laboratories system designers adopted the regular hexagonal shape several years ago [3].

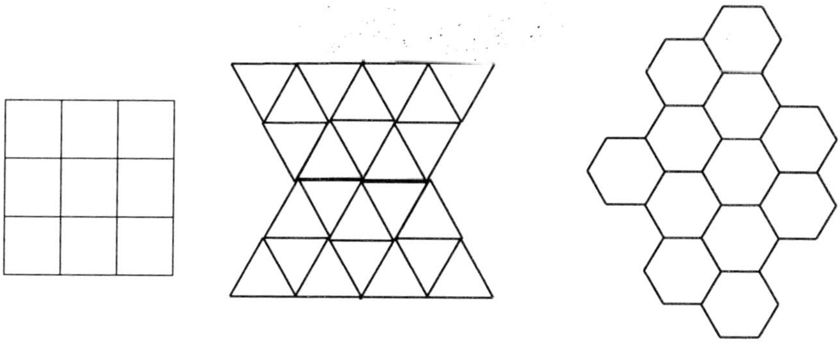

(a) (b) (c)

Fig.2: Regular polygon as cell. (a) Squares (b) Equilateral triangles (c) Regular hexagons

1.1.3 Function Description of System Operation

The main entities of an Advanced Mobile Phone Service system are the Mobile Telephone Switching Office (MTSO), the cell sites, and the mobile units. The central processor of the MTSO controls not only switching equipment needed to interconnect mobile parties with the land telephone network, but also cell-site actions and even many of the actions of mobile units through commands relayed to them by the cell sites.

The MTSO is linked with each cell site by a group of voice trunks-one trunk for each radio channel installed in the site-and two or more data links, over which the MTSO and cell site exchange information necessary for processing calls. Every cell site contains one transceiver for each voice channel assigned to it and the transmitting and receiving antennas for these channels. The cell site also contains signal-level monitoring equipment and a "setup" radio, whose purpose is explained below.

The mobile equipment consists of a control unit, a transceiver, a logic unit, and two antennas. The control unit contains all the user interfaces, such as a handset, various push buttons, and indicator lights. The transceiver uses a frequency synthesizer to tune to any allocated channel. The logic unit interprets customer actions and system commands and controls the transceiver and control units. A single antenna is used for transmission; two antennas together are used to provide space diversity for reception.

A few allocated radio channels serve as "setup" channels rather than voice channels. These channels are used primarily for the exchange of information needed to establish or set up calls. Applying the frequency-reuse concept to setup channels minimizes the number of channels withheld from voice use. Ordinarily, each site has one such channel. Whenever a mobile unit is turned on but the user is not engaged in a call, the mobile unit simply monitors a setup channel. The unit itself chooses which one of the various channels to monitor by sampling the signal strength on all members of a standard group setup channels. The mobile unit then tunes to the channel, which yields the strongest measurement, synchronizes with the data stream being transmitted by the system, and begins interpreting the data. Ordinary, the mobile unit will remain on this channel; in some cases, the received data will indicate that the mobile unit should sample the signal strength on another set of channels before making a final choice. The mobile unit continues to monitor the chosen setup channel unless some condition, such as poor reception, requires that the choice of a channel be renewed. The setup-channel data words include the identification numbers of mobile units to which calls are currently being directed.

When a mobile unit detects that it is being called, it quickly samples the signal strength on all the system's setup channels so that it can respond through the cell sites offering the strongest signal at the mobile unit's current position. The mobile unit seizes the newly chosen setup channel and transmits its page response. The system then transmits a voice-channel assignment addressed to the mobile unit, which, in turn, tunes to the assigned channel, where it receives a command to alert the mobile user. A similar sequence of actions takes place when the mobile user originates a call.

While a call is in progress, at intervals of a few seconds the system examines the signal being received at the serving call site (the site that is handling the call). When necessary, the system searches for another site to serve the call. When it finds a suitable site, the system sends the mobile unit a command to retune to a channel associated with that site. While the mobile unit is changing channels, the MTSO reswitches the land party to the trunk associated with the new channel's transceiver. The periodic examination of a mobile unit's signal is known as "locating." The act of changing channels has come to be called "hand-off" [3].

1.1.4 Hand-off Strategies

When a mobile moves into a different cell while a conversation is in progress, the Mobile Switching Center (MSC) automatically transfers the call to a new channel belonging to the new base station. This hand-off operation not only involves identifying a new base station, but also requires that the voice and control signals be allocated to channels associated with the new base station.

Processing hand-offs is an important task in any cellular radio system. Many hand-off strategies prioritize hand-off requests over call initiation requests when allocating unused channels in a cell site. Hand-off calls must be performed successfully and as infrequently as possible, and be imperceptible to the user.

In first generation analog cellular systems, signal strength measurements are made by the base stations and supervised by the MSC. Each base station constantly monitors the signal strengths of all its reverse voice channels to determine the relative location of each mobile user with respect to the base station tower.

In second generation systems that use digital Time Division Multiple Access (TDMA) technology, hand-off decisions are mobile assisted. In Mobile Assisted Hand-Off (MAHO), every mobile station measures the received power from surrounding base stations and continually reports the results of these measurements to the serving base station. A hand-off is initiated when the power received from the base station of neighboring cell begins to exceed the power received from the current base station by a certain level or for a certain period of time. The MAHO method enables the call to be handed over between base stations at a much faster rate than in first generation analog systems since the hand-off measurements are made by each mobile, and the MSC no longer constantly monitors signal strengths. MAHO is particularly suited for microcellular environments where hand-off calls are more frequent.

Different systems have different policies and methods for managing hand-off requests. Some systems handle hand-off requests in the same way they handle originating calls. In such systems, the probability that a hand-off request will not be served by a new base station is equal to the blocking probability of incoming calls. However, from the user's point of view, having a call abruptly terminated while in the middle of a conversation is more annoying than being blocked occasionally on a new call attempt. To improve the quality of service as perceived by the users, various methods have been devised to prioritize hand-off requests over call initiation requests when allocating voice channels [4].

1.1.5 Prioritizing Hand-off Calls

One method for giving priority to hand-off calls is called the guard channel concept, whereby a fraction of the total available channels in a cell is reserved exclusively for hand-off requests from ongoing calls, which may be handed off into the cell. This method has the disadvantage of reducing the total carried traffic, as fewer channels are allocated to originating calls. Guard channels, however, offer efficient spectrum utilization when dynamic channel assignment strategies, which minimize the number of required guard channels by efficient demand base allocation, are used.

Queuing of hand-off requests is another method to reduce the probability of forced termination of a call due to lack of available channels. There is a tradeoff between the decrease in probability of forced termination and total carried traffic. Queuing of hand-off calls is possible due to the fact that there is a finite time interval between the time the received signal level drops below the hand-off threshold and the time the call is terminated due to insufficient signal level. The delay time and size of the queue is terminated from traffic pattern of the particular service area. It should be noted that queuing does not guarantee a zero probability of forced termination, since large delays will cause the received signal level to drop below the minimum required level to maintain communication and hence lead to forced termination [4].

1.2 Analysis

Telecommunication systems play a major role in today's life more than ever. The new telecommunication services especially due to the number of users of the non-traditional services require a careful design of the network for optimal use of resources and providing a guaranteed quality of service to the users.

The growth of telecommunication system is more than ever stimulating the need for mathematical modeling and evaluation techniques to predict the performance of these systems. These techniques have become an indispensable tool for assessment of the performance of the system, evaluation of design alternatives, resource dimensioning, and system configuration.

Methods for performance evaluation fall into three categories: measurement, simulation, and analytical techniques. The measurement techniques have the advantage of direct measurement of network performance, which means no detail of network operation is excluded but it is expensive. Also, experimental network may be limited in the number and type of traffic sources available, thus restricting the range of realistic experimental conditions.

In comparing analytical and simulation techniques, the main factors to consider are the accuracy of the results, the time to produce results, and the overall cost of using the method (this includes cost of development as well as use). One advantage of analytical solution is that they can be used reasonably quickly. However, the need to be able to solve the model restricts the range of system or traffic characteristics that can be included. On the other hand, networks of almost arbitrary complexity can be investigated using simulation. However, simulation can be costly to develop and run, and it is time consuming, particularly when very rare events (such as cell loss and call blocked) are being measured. In contrast, analytical method for this case may be a quick and appropriate technique (inexpensive) [5].

The main idea behind the performance evaluation using analytical techniques depends on the fundamental of queuing theory. Queues naturally arise in telecommunication networks, whenever customers contend for limited shared resources. Customers (calls) will arrive in a special pattern to the queuing system and need a certain amount of service (waiting time) if the service is not available and then they are served and leave the system.

Any queuing system is described by arrival process, service process, the number of servers (channels), and the capacity of the system. For example, a queuing system with Poisson arrival and independent, identically distribution service time with infinite (finite) buffer and one server is known as a (finite) M/G/1 queue [9].

2 PROBLEM DEFINITION

2.1 The Cellular System Description

The cellular system we have studied has two types of calls: The new calls and the hand-off calls. For analytical analysis and performance evaluation of a cellular system, we have to model the events of the system with distributions that are able to capture the characteristics of those events. Poisson distribution is a good representative for most of the telecommunication events. In the former studies [1, 6, 10], Poisson distribution has been used to model the new calls and also the hand-off calls. The hand-off calls have different behavior compared to the new calls, so modeling both calls with Poisson distribution is not adequate. The hand-off calls have bursty characteristic which means that they are happening with different rates. The Poisson process is not able to capture the burstiness of the hand-off calls.

In this context, Markov Modulated Poisson Process (MMPP) is an n-state Markov chain that has different rates in each state and with more parameters compared with Poisson distribution. Therefore, it is able to capture diverse characteristics of the hand-off calls.

In the cellular system studied here, the hand-off calls have higher priority compared to the new calls. The hand-off calls are the ongoing calls when customers are moving from one cell to another, and they do not want to lose the connection in the middle of their conversation. The new calls are generated by customers who are trying to have access to an available channel. In case of unavailability of some free channels, the new calls will be buffered and be waiting for the next available channel.

In each cell, there are n channels (circuits or servers) to handle the incoming calls, among them g channels are reserved for hand-off calls (channels, circuits, and servers are used throughout the dissertation interchangeably). These g-reserved channels are called guard band channels. The reserved channels in each cell have been used to implement the prioritization.

An infinite buffer has been considered in the cellular system for the new calls' customer, which do not receive a channel to initiate a call. If n-g channels or more are available, the new call can have access to a channel and be processed; otherwise it will join the infinite queue. The new calls that are in the infinite queue will have access to the channel in a First Come First Serve (FCFS) discipline as soon as n-g or more channels will be available. Different customers require different amounts of service time. Hence, we have to use a probability distribution to describe the service pattern. The most commonly assumed service time distribution is the negative exponential distribution or, equivalently, the service processes are assumed to be Poisson process. In this cellular system, service rate for all calls are assumed to be Poisson distribution.

In the next section, we discuss how the hand-off calls were modeled in previous work then will present the new model for the hand-off calls.

2.2 The Hand-off Calls Modeling

2.2.1 Former Model

In most of the elementary queuing systems, the arrivals are assumed to be Poisson process. The reason behind Poisson's use lies in the fact that it possesses a number of marvelous probabilistic properties that give rise to many elegant queuing results. In the literature, Poisson distribution was used to modeling the hand-off calls. The new calls were also modeled with the Poisson process. The service times were assumed to be exponentially distributed or, equivalently, the service process was assumed to be Poisson process. The hand-off calls' rate was represented with λ, new calls' rate was represented with γ, and service rate was represented with μ.

2.2.2 New Model

The Poisson process has a limited number of parameters; therefore it is not always adequate in modeling the real time data traffic, which experience correlation in their arrivals. The hand-off call arrivals are in some sense correlated and exhibit a diverse mixture of characteristics with different correlation and burstiness. As queuing system performances are highly sensitive to these correlation and bursty characteristics in the arrival process, we need more realistic models to represent the arrival process.

Since, with the Poisson process model we were not able to take into consideration the bursty behavior of some of the incoming traffic due to the Poisson parameter's shortage; a new model with more parameters is necessary in order to overcome the drawback of the previous models [1, 6, 10].

In this research work, we have considered two states Markov Modulated Poisson Process (MMPP) to model the hand-off calls, but we have left the Poisson distribution for the new call arrivals modeling as well as the service process. Fig.3 shows the state diagram of the hand-off calls modeled with two-state MMPP.

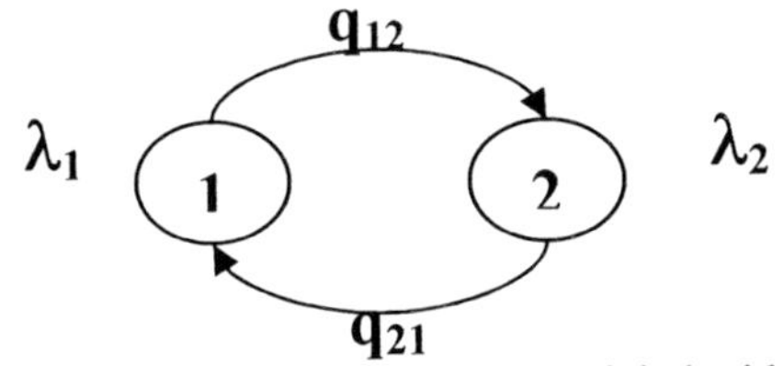

Fig.3: The state diagram of the hand-off calls modeled with two-state MMPP

Where;

State one: The state that the hand-off calls have λ_1 arrival rate

State two: The state that the hand-off calls have λ_2 arrival rate

λ_1: The hand-off calls arrival rate in state-one

λ_2: The hand-off calls arrival rate in state-two

q_{12}: Probability of transmission from the state-one (with rate λ_1) to the state two (with rate λ_2)

q_{21}: Probability of transmission from the state-two (with rate λ_2) to the state one (with rate λ_1)

In next chapter, we will present the solution process to find the performance measurements of both models (discussed in sections 2.2.1 and 2.2.2).

3 SOLUTION

3.1 Former Model

Performance evaluation aims at forecasting system behavior in a quantitative way. By modeling the events in a system, we will be able to analyze the system and calculate different performance measurements. We tried to calculate performance measurements such as the probability of blocking, the queue length distribution, and the average queuing time by solving the cellular system for both the Poisson and MMPP models of the hand-off calls. We solved the former model (the hand-off calls modeled with Poisson process) in the same way as we solved the new model (the hand-off calls modeled with MMPP). We were able to determine the desired performance measurements for both models because we followed the same procedure to solve both models. The next step was to compare between their performance measurements computed for both models and see the difference between them. The differences in results between models' performance measurements were impacted from bursty characteristics of the hand-off calls, which were captured with MMPP modeling.

In the next section, we will present the state diagram and transition matrix for former model [1].

3.1.1 State of the Cellular System

In our system, each of the cell is assumed to have *n* channels, *g* of these channels have been reserved for hand-off calls. If *n*-*g* or more channels in a cell are busy at the instant, a new call arrives, then this call is joined to infinite buffer (queue). If all n channels in a cell are busy at the instant a hand-off call arrives, this call is blocked and is assumed to be lost from the system. Such a network can be modeled as a two-dimension continuous–time Markov chain. The behavior of the cellular system can be described by using two-state Morkovian chain as follows: $S(i, j)$ where;

> i: Level of the system that shows the number of new calls in the infinite queue
> j: Phase of the system, which shows the numbers of busy servers
> $i \geq 0$ because the buffer size is an infinite, any number of new calls can be in the queue
> $0 \leq j \leq$ n, j is limited to the number of servers (channels) in each cell

3.1.2 State Diagram

Fig.4 shows the state diagram for the Poisson modeled hand-off calls' cellular system.

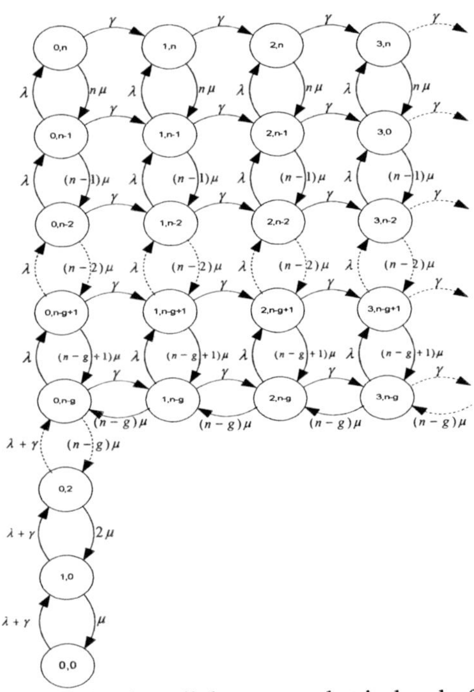

Fig.4: State diagram for the cellular system that its hand-off calls are modeled with Poisson.

3.1.3 Transition Matrix

In order to be able to study the performance of our system, we need to consider the transition matrix that is going to represent all transition probabilities among the states of the whole state diagram. After considering transition probabilities between the states of the continuous time Markov chain (shown in fig.4) and partitioning it into blocks of matrices, the transition matrix Q will be as follows:

$$Q = \begin{vmatrix} B_{00} & B_{01} & 0 & 0 & 0 & 0 & \cdots \\ B_{10} & B_{11} & A_0 & 0 & 0 & 0 & \cdots \\ 0 & B_{20} & A_1 & A_0 & 0 & 0 & \cdots \\ 0 & 0 & A_2 & A_1 & A_0 & 0 & \cdots \\ 0 & 0 & 0 & A_2 & A_1 & A_0 & \cdots \\ . & . & . & . & . & . & \cdots \\ . & . & . & . & . & . & \cdots \end{vmatrix}$$

The matrices B_{00} and B_{11} represent transitions among levels zero and one, respectively. These are square matrices with size *(n-g)* × *(n-g)* and *(g+1)* × *(g+1)*, respectively.

$$B_{00} = \begin{vmatrix} -(\lambda+\gamma) & \lambda+\gamma & \cdots & 0 \\ \mu & -(\lambda+\gamma+\mu) & \cdots & 0 \\ . & . & \cdots & . \\ . & . & \cdots & . \\ 0 & . & \cdots & \lambda+\gamma \\ 0 & . & \cdots & -(\lambda+\gamma+(n-g-1)\mu) \end{vmatrix}$$

$$B_{11} = \begin{vmatrix} -(\lambda+\gamma+(n-g)\mu) & \cdots & 0 \\ (n-g+1)\mu & \cdots & 0 \\ . & \cdots & . \\ . & \cdots & . \\ 0 & \cdots & \gamma \\ 0 & \cdots & -(\gamma+n\mu) \end{vmatrix}$$

The matrices B_{10} and B_{01} represent the transitions from level one to level zero and vise versa. Their dimensions are $(g+1) \times (n\text{-}g)$ and $(n\text{-}g) \times (g+1)$, respectively. The matrix B_{20} represents transition from level two to level zero; its dimension is $(g+1) \times (g+1)$. B_{20} and B_{01} are computed as follows:

$$B_{10} = \begin{vmatrix} 0 & . & . & 0 & (n-g)\mu \\ 0 & . & . & . & 0 \\ . & . & . & . & . \\ . & . & . & . & . \\ . & . & . & . & . \\ 0 & . & . & . & 0 \end{vmatrix} \quad B_{20} = \begin{vmatrix} (n-g)\mu & . & . & . & 0 \\ 0 & . & . & . & 0 \\ . & . & . & . & . \\ . & . & . & . & . \\ . & . & . & . & . \\ 0 & . & . & . & 0 \end{vmatrix} \quad B_{01} = \begin{vmatrix} 0 & . & . & . & 0 \\ 0 & . & . & . & . \\ . & . & . & . & . \\ . & . & . & . & . \\ . & . & . & . & . \\ \lambda+\gamma & . & . & . & 0 \end{vmatrix}$$

The matrix A_0 is a $(g+1)$ square diagonal matrix that represents transitions from level $(i\text{-}1)$ to level i for $i > 2$; $A_0 = diag\ (\gamma, \gamma, \gamma, \ \dots\ \gamma)$.

$$A_0 = \begin{vmatrix} \gamma & 0 & \dots & 0 \\ 0 & \gamma & \dots & 0 \\ . & . & . & . \\ 0 & \dots & \dots & \gamma \end{vmatrix}$$

The remainder matrices are A_1 and A_2. They are computed as: $A_1 = B_{11}$ and $A_2 = B_{20}$. In the next section, we will present the state diagram and transition matrix for the system modeled with MMPP.

3.2 Markov Modulated Poisson Process (MMPP) Model for Hand-off Calls

The hand-off calls are modeled with a two-state Markov Modulated Poisson Process to find out the effect of bursty characteristics on the performance measurements. The new call arrival rate and the service rate of the cellular system are assumed to have Poisson distribution. All calls are serviced at rate μ according to a Poisson process. The new calls arrive to a given cell according to the Poisson process with rate γ. A mobile user moves from one cell to another according to a two-state MMPP. Two-state MMPP have two different arrival rates in each of the states. The hand-off calls arrival rate in state-one is λ_1 and their arrival rate in state-two is λ_2. In two-state MMPP, entering any state and leaving that state have different probabilities. q_{12} is the probability of leaving the state-one and entering state-two, and q_{21} is the probability of leaving state-two and entering state-one. Using the following equation (1), the average arrival rate for hand-off calls can be determined:

$$\bar{\lambda} = \frac{q_{21}}{q_{12} + q_{21}} \lambda_1 + \frac{q_{12}}{q_{12} + q_{21}} \lambda_2 \qquad (1)$$

In Poisson modeling, we just have an arrival rate for introducing the hand-off calls. In a two-state MMPP model, we have four different parameters to introduce the hand-off calls and can determine an average arrival rate. In order to be able to determine the impact of the bursty behavior of the hand-off calls on the performance metrics, one assumption has to be made. This assumption consists on equalizing the hand-off calls' arrival rate of the Poisson model and the average hand-off calls' arrival rate of the MMPP model.

In the following section, we present the cellular system where hand-off calls are modeled with two-state MMPP. Finally, we provide a solution to quantify the performance measurements.

3.2.1 State of the Cellular System

In our study, the cellular system with same characteristics has been used. However, the hand-off calls are modeled with a two state MMPP instead of Poisson process. The cellular system characteristics are as follow: each cell is assumed to have n channels, g of these channels has been reserved for hand-off calls. If n-g or more channels in a cell are busy at the instant a new call arrives, then this call joins an infinite buffer (queue). If all n circuits in a cell are busy at the instant a hand-off call arrives, this call is blocked and assumed lost from the system. As we saw in section 3.1.1, such a system can be modeled as a two-dimension continuous–time Markov chain, but we have to consider a two-state MMPP that has been used to model the hand-off calls.

We need to add one more phase to two-dimension continuous-time Markov chain to represent both: the cellular system and the two-state of

MMPP. Now we have a three-dimensional continuous-time Markov chain that can be described as follows: S (i, j, k)

Where;

i: is the number of the new call customers in the infinite queue
j: is the number of busy circuits (channels) that are carrying the calls in a cell
k: is representing the state of MMPP
$i \geq 0$ because we have assumed an infinite buffer for new calls to join, $0 \geq j \geq n$ because the number of busy circuits are limited to the number of available circuits (channels) in a cell, and k can be either one or two, which are two states of MMPP that have been used to model the hand-off calls.

3.2.2 State Diagram

The state diagram represents the behavior of the cellular system. Having a three-dimensional continuous Markov chain, the state diagram will be three-dimensional too. There are two, two-dimensional state diagrams that are connected to each other in every single state. Figure 5 and 6 represent two-dimensional state diagrams with a hand-off call's rate λ_1 and λ_2 respectively. Figure 7 represents a three-dimensional state diagram which, was obtained with combining two two-dimensional state diagram into two layers: one front layer and one back (shaded) layer. The transition from one layer to another represents the transition probability between two states of MMPP model.

When the number of busy channels in a cell is less than n-g channels, all the arrived new calls and hand-off calls are serviced and there will be no queuing. Part (a) of figure 5 and 6 shows this situation. Whenever the number of busy channels in a cell is greater than n-g, only the arrived hand-off calls will be serviced. This situation is shown in the vertical portion of figure 5 and 6 part (b). In case of more than n hand-off call arrival only n calls will be serviced however, the remaining calls will be blocked. The horizontal portion of part (b) in the figure 5 and 6 shows the new call arrival with rate γ which, are joining the infinite queue. Figure 8 shows the transition probability between the two layers of three-dimensional continuous time Markov chain's state diagram represented in figure 7.

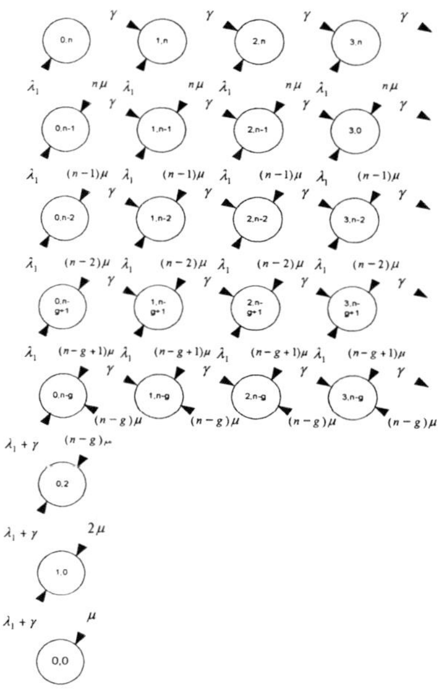

Fig.5

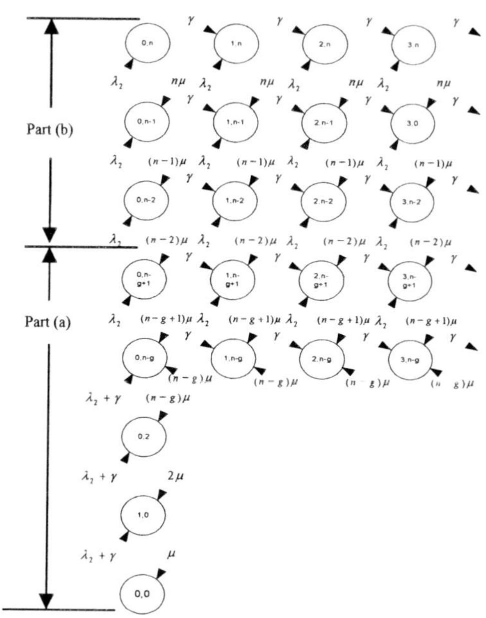

Fig.6

Figure 5: the state diagram of the two-dimensional continuous-time Markov chain with λ_1 as the hand- off calls' arrival rate.

Figure 6: the state diagram of the two-dimensional continuous-time Markov chain with λ_2 as the hand-off calls' arrival rate.

ed. note: need to paste in figure 8

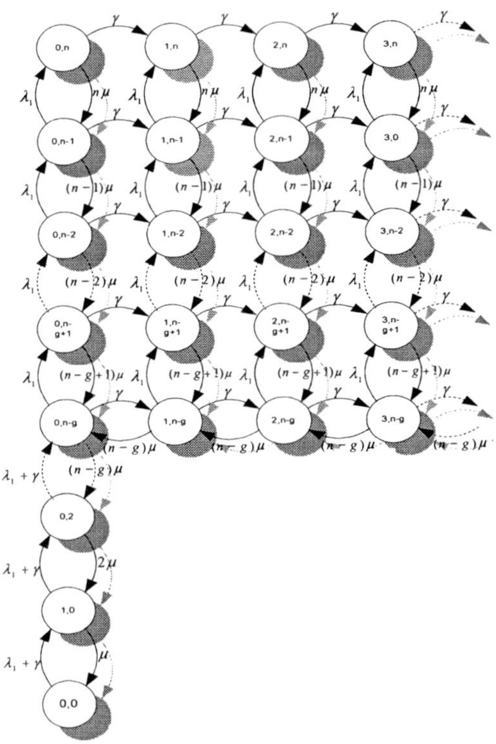

Fig.7

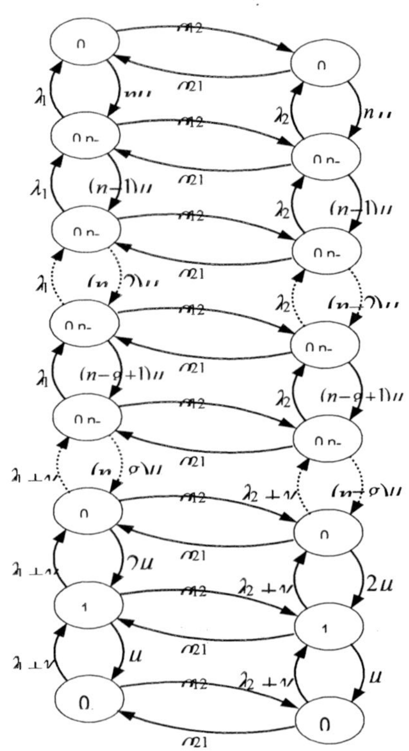

Fig.8

Fig. 7: the three-dimensional continuous-time Markov chain's state diagram.

Fig. 8: a portion of the side view of the three-dimensional continuous-time Markov chain's state diagram

3.2.3 Transition Matrix

The transition matrix can be derived from the state diagram in figure 7 and it will form a transition matrix similar to that of the Poisson model. Transition matrix Q shown below can be derived from the three-dimensional state diagram in figure 7 section 3.2.2.

$$Q = \begin{vmatrix} B_{00} & B_{01} & 0 & 0 & 0 & 0 & \ldots\ldots \\ B_{10} & B_{11} & A_0 & 0 & 0 & 0 & \ldots\ldots \\ 0 & B_{20} & A_1 & A_0 & 0 & 0 & \ldots\ldots \\ 0 & 0 & A_2 & A_1 & A_0 & 0 & \ldots\ldots \\ 0 & 0 & 0 & A_2 & A_1 & A_0 & \ldots\ldots \\ . & . & . & . & . & . & \ldots\ldots \\ . & . & . & . & . & . & \ldots\ldots \end{vmatrix}$$

However, each of the elements has been affected by two-state MMPP. Every single scalar member of the Poisson model's transition matrix will change to 2×2 square matrix. Therefore, the overall size of the transition matrix becomes twice as much as it was in the Poisson model.

The sizes of the blocks inside the transition matrix also change as follows:

The matrix B_{00}'s dimension was $(n\text{-}g) \times (n\text{-}g)$ and it will be $2(n\text{-}g) \times 2(n\text{-}g)$.
The matrix B_{01}'s dimension was $(n\text{-}g) \times (g+1)$ and it will be $2(n\text{-}g) \times 2(g+1)$.
The matrix B_{10}'s dimension was $(g+1) \times (n\text{-}g)$ and it will be $2(g+1) \times 2(n\text{-}g)$.
The matrices A_{11}, A_0, A_2, B_{21}, and B_{11}'s dimensions were $(g+1) \times (g+1)$ and they will be $2(g+1) \times 2(g+1)$.

Each element in each of the matrices will change according to the following matrices:

$$-(\lambda+\gamma+(n-g-i)\mu) = \begin{vmatrix} -(\gamma+(n-g-i)\mu+\lambda_1+q_{12}) & q_{12} \\ q_{21} & -(\gamma+(n-g-i)\mu+\lambda_2+q_{21}) \end{vmatrix}$$

For $1 \le i \le n\text{-}g$

$$(n-i)\mu = \begin{vmatrix} (n-i)\mu & 0 \\ 0 & (n-i)\mu \end{vmatrix}$$

For $0 \le i \le g\text{-}1$

For $0 \le i \le n\text{-}g\text{-}1$

$$-(\lambda+\gamma+(n-i)\mu) = \begin{vmatrix} -(\gamma+(n-i)\mu+\lambda_1+q_{12}) & q_{12} \\ q_{21} & -(\gamma+(n-i)\mu+\lambda_2+q_{21}) \end{vmatrix}$$

$$(n-g-i)\mu = \begin{vmatrix} (n-g-i)\mu & 0 \\ 0 & (n-g-i)\mu \end{vmatrix}$$

For $0 \le i \le g$

$$\gamma = \begin{vmatrix} \gamma & 0 \\ 0 & \gamma \end{vmatrix} \qquad \lambda = \begin{vmatrix} \lambda_1 & 0 \\ 0 & \lambda_2 \end{vmatrix}$$

The matrices B_{00} and B_{11} represent transitions among the levels zero and one respectively. The matrices B_{10} and B_{01} represents transitions from level one to level zero and vise versa. The matrix B_{20} represents a transition from level two to level zero. All these matrices are shown as follows:

$$B_{21} = \begin{vmatrix} (n-g)\mu & 0 & \cdots & 0 & 0 \\ 0 & (n-g)\mu & \cdots & 0 & 0 \\ 0 & 0 & & & \\ 0 & 0 & \cdots & & \\ \vdots & & \ddots & & \vdots \\ 0 & 0 & \cdots & 0 & 0 \\ 0 & 0 & \cdots & 0 & 0 \end{vmatrix} \quad B_{01} = \begin{vmatrix} 0 & 0 & \cdots & 0 & 0 \\ 0 & 0 & \cdots & 0 & 0 \\ \vdots & & \ddots & & \vdots \\ \lambda_1+\gamma & 0 & \cdots & 0 & 0 \\ 0 & \lambda_2+\gamma & \cdots & 0 & 0 \end{vmatrix} \quad B_{10} = \begin{vmatrix} 0 & 0 & \cdots & (n-g)\mu & 0 \\ 0 & 0 & \cdots & 0 & (n-g)\mu \\ 0 & 0 & & 0 & 0 \\ 0 & 0 & \cdots & 0 & 0 \\ \vdots & & \ddots & & \vdots \\ 0 & 0 & \cdots & 0 & 0 \\ 0 & 0 & \cdots & 0 & 0 \end{vmatrix}$$

$$B_{00} = \begin{vmatrix} -(q_{11}) & q_{12} & \lambda_1+\gamma & 0 & \cdots & 0 & 0 \\ q_{21} & -(q_{22}) & 0 & \lambda_2+\gamma & \cdots & 0 & 0 \\ \mu & 0 & -(q_{11}+\mu) & q_{12} & & 0 & 0 \\ 0 & \mu & q_{21} & -(q_{22}+\mu) & \cdots & 0 & 0 \\ \vdots & & \vdots & & \ddots & \vdots & \\ 0 & 0 & & & \cdots & \lambda_1+\gamma & 0 \\ 0 & 0 & & & \cdots & 0 & \lambda_2+\gamma \\ 0 & 0 & & & & -(q_{11}+(n-g-1)\mu) & q_{12} \\ 0 & 0 & & & \cdots & q_{21} & -(q_{22}+(n-g-1)\mu) \end{vmatrix}$$

Where;

$$q_{11} = \lambda_1 + q_{12} + \gamma$$
$$q_{22} = \lambda_2 + q_{21} + \gamma$$

$$B_{11} = \left|\begin{array}{ccccc}
-(\lambda_1 + q_{12} + \gamma + (n-g)\mu) & q_{12} & \cdots & 0 & 0 \\
q_{21} & -(\lambda_2 + q_{21} + \gamma + (n-g)\mu) & \cdots & 0 & 0 \\
(n-g+1)\mu & 0 & & 0 & 0 \\
0 & (n-g+1)\mu & \cdots & 0 & 0 \\
\vdots & & \cdots & \vdots & \\
0 & 0 & \cdots & \gamma & 0 \\
0 & 0 & \cdots & 0 & \gamma \\
0 & 0 & & -(\gamma + n\mu) & 0 \\
0 & 0 & \cdots & 0 & -(\gamma + n\mu)
\end{array}\right|$$

The matrix A_0 represents a transition from level $(i\text{-}1)$ to level i for $i \geq 2$.

$$A_0 = \left|\begin{array}{cccccc}
\gamma & 0 & \cdots & \cdots & 0 & 0 \\
0 & \gamma & & & 0 & 0 \\
0 & 0 & \gamma & 0 & 0 & 0 \\
0 & 0 & 0 & \gamma & \cdots & 0 \\
\vdots & & \vdots & \vdots & & \vdots \\
0 & 0 & & & \gamma & 0 \\
0 & 0 & \cdots & \cdots & 0 & \gamma
\end{array}\right|$$

The remainder matrices are A_1 and A_2 such that: $A_1 = B_{11}$ and $A_2 = B_{20}$.

In the next section, we present a solution for continuous time Markov chain with transition matrix Q that was shown in section 3.2.3 and 3.1.3.

3.2.4 Solving Quasi Birth and Death (QBD) Process with Complex Boundary

The generator matrix Q in section 3.1.3 has the form of a Quasi Birth and Death (QBD) process with complex boundary. A birth–death process is a special case of Markov chain in which the process makes transitions only to neighboring states with respect to its current position. Matrices A_0, A_1, and A_2 represent the birth and death process and B_{00}, B_{11}, B_{01}, and B_{10}, B_{20} form the complex boundary. The matrix A_1 has only transitions to one level after (birth) and one level before (death) of its position.

Using Nuets' method [2] we can solve a QBD with complex boundary. A QBD with complex boundary has a matrix–geometric solution. Marcel F. Nuets pioneered this particular method of solving queuing systems. It takes advantage of the similar structure presented in many interesting stochastic models and formulates its solution in terms of a non-linear matrix equation.

The steady–state vector of Markov chain $\pi = (\pi_0, \pi_1, \pi_2, \pi_3 ...)$ can be found by solving the following equations.

$$\begin{cases} \pi Q = 0 \\ \sum_{i=0}^{\infty} \pi_i e = 1 \end{cases}$$

e is a column vector of ones with a proper size.

The first equation is equivalent to the following set of equations:

$$\begin{cases} \pi_{01} B_{00} + \pi_{02} B_{10} = 0 \rightarrow (2) \\ \pi_{01} B_{01} + \pi_{02} B_{11} + \pi B_{21} = 0 \rightarrow (3) \\ \pi_{02} A_0 + \pi_1 A_1 + \pi_2 A_2 = 0 \rightarrow (4) \\ \pi_{i-1} A_0 + \pi_i A_1 + \pi_{i+1} A_2 = 0 \ for \ i \geq 2 \rightarrow (5) \end{cases}$$

From equation (4) and (5), we can get (see [2] for more details):

$$\pi_i = \pi_{02} R^i \ for \ i \geq 1 \rightarrow (6)$$

Where R satisfies:

$$A_0 + RA_1 + R^2 A_2 = 0 \rightarrow (7)$$

Equation (7) can be solved for R by using recursion with $R_0 = 0$;

$$R_j = -(A_0 A_1^{-1} + R_{j-1A}{}^2 A_2 A_1^{-1})\ for\ i \geq 1 \rightarrow (8)$$

Equation (2) and (3) with respect to equation (6) can be rewritten as follow:

$$\begin{cases} \pi_{01} B_{00} + \pi_{02} B_{10} = 0 \\ \pi_{01} B_{01} + \pi_{02} B_{11} + \pi_{02} R B_{21} = 0 \end{cases}$$

Or we can write the joint equations as follows:

$$|\ \pi_{01} \quad \pi_{02}\ \left| \begin{vmatrix} B_{00} & B_{01} \\ B_{10} & B_{11} + RB_{21} \end{vmatrix} \right. = 0$$

Where another matrix $B[R]$ can be computed:

$$B[R] = \begin{vmatrix} B_{00} & B_{01} \\ B_{10} & B_{11} + RB_{21} \end{vmatrix}$$

$B[R]$ is a stochastic matrix which means the sum of each of its rows is zero. Therefore, it has a non-zero solution.

$$\pi_0 = |\pi_{01} \quad \pi_{02}|$$

π_0 is a boundary vector and can be computed by solving the following equations:

$\pi_0\, B[R] = 0$ and $B[R]\, e = 0$

Finally, the following equation has to be satisfied for normalization:

$$\pi_{01}e + \pi_{02}e + \sum_{i=1}^{\infty} \pi_i e = 1 \rightarrow (9)$$

We have a closed form for summation as:

$$\sum_{j=1}^{\infty} X^j = X(X-1)^{-1} \rightarrow (10)$$

By considering equation (6) and (10) we can write the summation in equation (9) as follows:

$$\sum_{i=1}^{\infty} \pi_i e = \sum_{i=1}^{\infty} \pi_{02} R^i e = \pi_{02} R(I-R)^{-1} e$$

Therefore, equation (8) can be rewritten as:

$$\pi_{01}e + \pi_{02}e + \pi_{02}R(I-R)^{-1}e = 1 \rightarrow (11)$$

By replacing π_{01}, π_{02} (found in vector boundary π_0 and R matrix in left side of the equation (11)), we get a positive number that divides π_{01}, π_{02}. The result of this division is to normalize π_{01}, π_{02}. When we replace the normalized π_{02} and R in equation (6), Stationary Probability of being in state i (π_i) will be found. In the next section, some of the performance measurements will be introduced and calculated.

3.2.5 Finding the Performance Measurements

π_i means the stationary probability of being in level i with no respect to phases. In other words, it is considered as the accumulated probabilities of being in different phases at a single level. The level represents the number of new calls in the infinite queue. However the phase represents the number of busy channels in each cell. With respect to the meaning of level and phase, the stationary probability of being in level i meaning the probability of having i new calls in the infinite queue without any concern about how many cell channels are busy. By having the stationary probability, we are easily able to find all performance measurements.

3.2.5.1 Stability Condition for the System

In queuing theory, the system has to be in the steady state for analytical analysis. For the system to be stable, the average drift to the right has to be smaller than the average drift to the left. In QBD model, $\pi A_0 e$ represents the average drift to the right and $\pi A_2 e$ represents average drift to the left. The system will be stable if the following condition is satisfied: $\pi A_2 e \geq \pi A_0 e$.

If in any case the system is not stable, we cannot trust the results that will be found from an unstable system.

In section 3.1, we presented the cellular system with Poisson modeling for the hand-off calls. In the next section, we will present the same cellular system but with two-state MMPP modeling for the hand-off calls.

3.2.5.2 Blocking Probability

The blocking probability is the probability that a hand-off call will be lost (once all channels are busy). The probability of blocking can be determined by adding the stationary probability of being in all different levels and the phase will be a maximum number of available channels in a cell *(n)*. In other words, blocking probability is the sum of the stationary probabilities for any number of the new calls in the queue when all channels are busy.

$$\pi_{01} = \left| \pi_{(0,1)} \quad \pi_{(0,2)} \quad \cdots\cdots\cdots\cdots \quad \pi_{(0,n-g-1)} \right|$$

$$\pi_{02} = \left| \pi_{(0,n-g)} \quad \pi_{(0,n-g+1)} \quad \cdots\cdots\cdots \quad \pi_{(0,n)} \right|$$

Expanded π_{01}, π_{02} and π_i are presented to observe their phases.

$$\pi_i = \left| \pi_{(i,n-g)} \quad \pi_{(i,n-g+1)} \quad \cdots\cdots\cdots \quad \pi_{(i,n)} \right| \; for\, i \geq 1$$

By considering the stationary probabilities, we can write the probability of

$$Blocking\ probability = \left(\pi_{(02,n)} + \pi_{(1,n)} + \pi_{(2,n)} + \pi_{(3,n)} + \cdots\cdots\cdots\right)$$

$$Blocking\ probability = \pi_{(02,n)} + \sum_{i=1}^{\infty} \pi_{(i,n)}$$

blocking as follows:

n = Maximum number of available channels in each cell

3.2.5.3 The New Calls Distribution

The probability of having a different number of new calls in the infinite queue is represented with stationary probabilities. Distribution of the new calls in the queue can be shown by plotting the stationary probability of having different new calls in the queue versus the different number of new calls in the queue. The new call distribution shows how they are populated in the infinite buffer.

3.2.5.4 The New Calls Average Queue Length

By having stationary probabilities, the average queue length can be determined. The Average queue length is the same as the mean of the queue distribution. In order to find average queue length we can use the following equation:

We know from equation (6) in section 3.1.4 that:

$$\pi_i = \pi_{02} R^i \;\; for\ i \geq 1 \rightarrow (6)$$

And also we know that:

$$\sum_{i=0}^{\infty} iR^i = R(I-R)^{-2}$$

Therefore, the average queue length will be:

Replacing π_{02} and R (known from solving QBD in section 3.2.4), we could calculate the average queue length represented by equation 12.

$$Average\ queue\ length = \sum_{i=0}^{\infty} i\pi_{02}R^i = \pi_{02}\sum_{i=0}^{\infty} iR^i = \pi_{02}(R(I-R)^{-2}) \rightarrow (12)$$

3.2.5.5 The New Calls Waiting Time in the Queue

Little's theorem relates the average number of customers (N) in a steady-state queuing system to the product of average arrival rate (λ) of customers entering the system and the average time (T) a customer spends in the system as follows:

$N=\lambda T$

This result was derived under very general conditions. The beauty of it lies in the fact that it does not assume any specific distribution for the arrivals as well as the service process, nor does it assume any queuing discipline or depend upon the number of parallel servers in the system. We have already computed average queue length in section 3.2.5.4.

$Average\ queue\ length = \pi_{02}\,(R\,(I\text{-}R)^{-2})$

Using Little's theorem, an average of waiting time in the queue (T_q) can be found as follows:

$$T_q = \frac{\pi_{02}(R(I-R)^{-2})}{\gamma}$$

The average of the total waiting time (T_w) is the sum of the service time (T_S) and average waiting time in the queue. This can be manifested in equations as:

$T_w = T_q + T_S$

$$T_S = \frac{1}{\mu}$$

Therefore total waiting time in the queue is:

$$T_w = \frac{\pi_{02}(R(I-R)^{-2})}{\gamma} + \frac{1}{\mu}$$

In the next section numerical results will be presented.

4 Numerical Results

In this chapter, analytical calculations will be performed using the computational procedures described in sections 3.2.4 and 3.2.5. By programming these formulas using Mathlab, we were able to plot some of the performance metrics under different conditions for both models of the hand-off calls. In this section, we will present each of the plots with a description of its assumptions and behaviors commenting on comparisons and contrasts, whenever apparent.

4.1 The Maximum Number of the Guard Band Channels for the Stable System

Figure 9 is plotted with the following assumptions: the arrival rate of the hand-off calls in state one of two-state MMPP is 5 calls per unit, however in state two is 1/5 calls per unit. The average arrival rate of the hand-off calls is 4.52, which is calculated using equation (1) in section 3.2. The arrival rate of the new calls is 1/5 calls per unit for MMPP and Poisson models. The arrival rate of the hand-off calls modeled using Poisson distribution is equal to the average arrival rate of the hand-off calls modeled by MMPP.

The figure also shows the areas that cellular system is stable with respect to the number of channels in a cell and the number of guard band channels. As we can observe, there are gaps between the curves calculated for both models. Representing the border between the stable and unstable areas is very important and has to be precise. Any small mistake to determine the stability region of the system will cause a wrong calculation for any performance measurements. The difference between these models is caused by the bursty analysis. Hence, we have to rely on the curve that is calculated using the MMPP model because MMPP has captured the specific characteristic of the hand-off calls. All the assumptions for plotting the future figures are made with respect to stability condition of the system.

4.2 The New Calls' Queue Distribution for Different Hand-off Calls' Arrival Rate

Figure 10 shows the new calls queue distribution for the hand-off calls, which are modeled with MMPP and Poisson process. The number of channels in a given cell is assumed to be 15 and among them 6 channels are assumed to be guard band channels. The new calls queue distribution for MMPP modeling is shown as a red color, however the green color is used for Poisson distribution modeling.

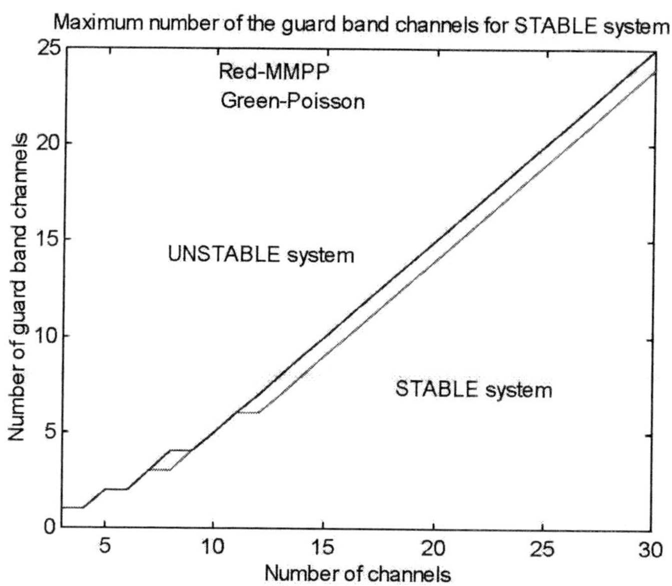

Fig.9: The maximum number of the guard band channels for a stable system

The queue distribution for both models is plotted when the arrival rate of the hand-off calls is 23 times, then 32 times the arrival rate of the new calls. In figure 10, the ratio represents the hand-off calls' arrival rate over new calls' arrival rate. From the same figure, we noticed that when the hand-off calls rate to the new calls rate ratio increases, the queue distribution for both models increases too. When the hand-off calls arrive to a cell with a higher rate than the new calls, they will occupy more channels.

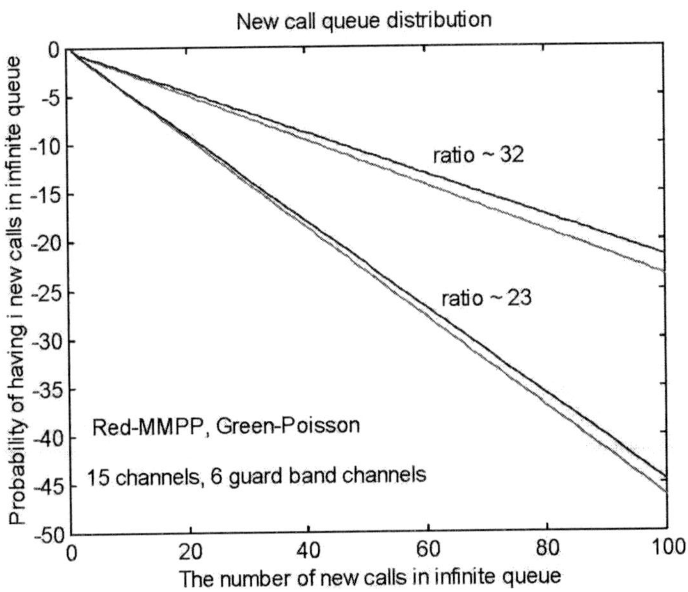

Fig.10: New calls' queue distribution for different hand-off calls' arrival rate

The new calls join the infinite queue while waiting for some channels to become free; therefore the queue distribution increases. Another thing to notice in the same figure, which consist as the following: there are some gaps between the curves of the two models. When the hand-off calls rate is 32 times the new calls rate, the gap between MMPP model and Poisson model is higher than when the hand-off calls rate to the new calls rate ratio is 23. As the ratio increases, the difference between the curves of both models becomes higher. When the hand-off calls rate increases, its bursty behavior will have more affect on the new calls' queue distribution. According to this first result, the MMPP model provided more accurate results than the Poisson model. This is due to the fact that MMPP has much more parameters for modeling incoming traffics than its counterpart, Poisson.

4.3 The New Call's Queue Distribution for Different Guard Band Channels

Figure 11 is plotted with the following assumptions: MMPP modeled hand calls' arrival rate in state one is assumed to be 5 calls per unit, and their arrival rate in state two is assumed to be 1/5 calls per unit. We would like to see the effect of the hand-off calls' arrival rate in an extreme case, therefore the hand-off calls arrival rate in two different states are assumed to be far apart. Under these assumptions, the average hand-off calls arrival rate given by equation (1) in section 3.2 will be 4.52 calls per unit. Poisson modeled hand-off calls arrival rate is assumed to be the same as the average hand-off calls arrival rate modeled by MMPP. The arrival rate of the new calls for both models is assumed to be 1/5 calls per unit. The number of channels available in a given cell is assumed to be 15. The new calls queue distribution for following guard band values: 2, 4, and 6. The curves for Poisson modeling were shown with green color, however the red color was used for MMPP modeling curves.

According to figure 4.3, for both models, as the guard band channels increase, the new calls queue distribution also increases. The reason is that when we reserve more channels for the hand-off calls, the new calls arrival will not access the channel easily, therefore they have to wait in the infinite queue for the next available channel.

Though changing the number of reserved channels have the same effect on both models, there are some gaps between red color and green color curves. Two different models used for the hand-off calls, which have caused a remarkable difference between these two cases (MMPP and Poisson). The Poisson model was not able to capture the real behavior of the hand-off calls, therefore the results of the analytical analysis would not be precise and reliable. The MMPP model has again provided more accurate results in this case compared to the Poisson model.

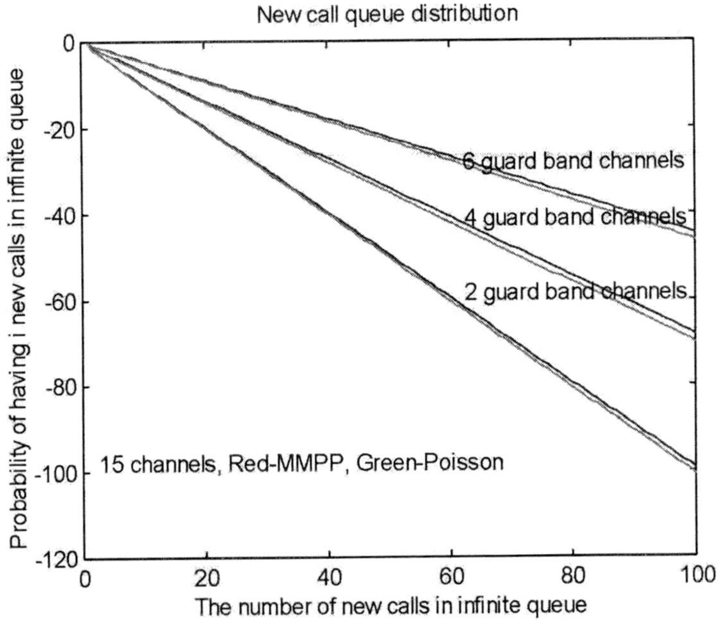

Fig.11: New calls' queue distribution for different guard band channels

4.4 Average New Calls' Queue Length versus the Number of Channels Available in a Cell

For both models, the new call arrival rate is assumed to be 1/5 calls per unit time. MMPP's curves are shown with the red color and Poisson's curves are shown with the green color. In this plot, three issues are captured: First, we consider the average new calls in the queue for both models versus the number of servers for three guard band channels.

As we can see from figure 12, for these three-guard band channels, when the number of channels in a cell increases, the average new calls in the queue decreases. Whenever the number of channels in a cell increases, the new call arrival will access the channels without being queued. Second, the same figure shows the increase of the number of guard bands' effect on the average new calls within the infinite queue. We observe for both models (Poisson and MMPP) that as the number of guard bands increase the average new calls in the queue also increases. When we have more guard bands in a cell, more channels are reserved for the hand-off calls, therefore, there is less chance for the arrived new calls to be serviced. The arrival new calls have to join the infinite queue and wait for an available channel, which causes an increase in the average number of new calls in the queue. Third, there are gaps between MMPP's curve and the Poisson's curves. These differences are caused by bursty characteristics of the handoff calls. As the number of guard band channels increases, the difference between two curves increases. For instance, when there are 15 servers and 6 guard band channels, the gap between the average new calls in the queue for both models is more important compared to that we are considering 2 guard band channels. The effect of the hand-off calls' bursty behavior increases when the number of guard bands increases.

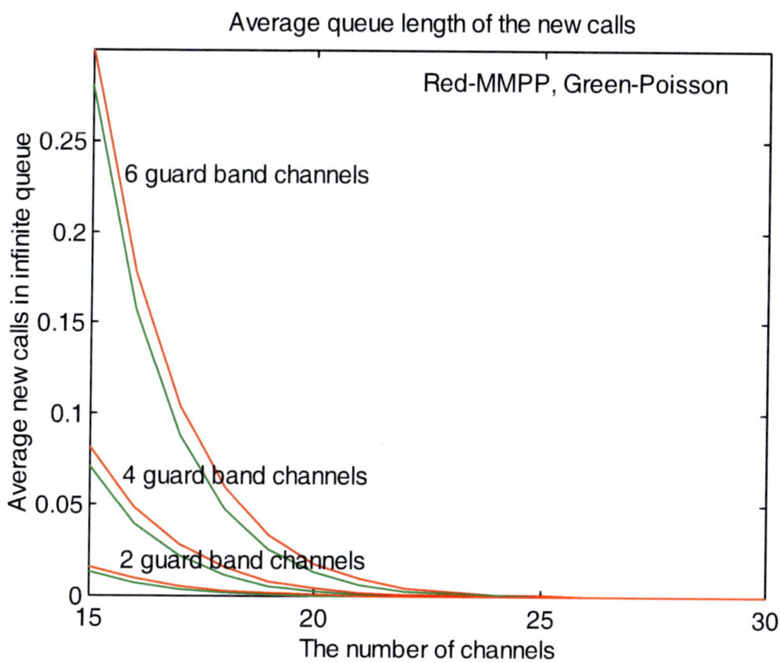

Fig.12: Average new calls' queue length vs. the number of available channel in a given cell for different guard band channels

4.5 The Average Hand-off Calls Rate Versus the Blocking Probability of the Hand-off Calls

In figure 13, we have assumed that the number of channels in a given cell is 15. For each model of the hand-off calls, two different numbers of guard band channels were considered. The curves corresponding to the 8 guard band channels was presented with green color, however we used the red color to represent that of the 2 guard band channels.

When the average arrival rate of the hand-off calls increases, the blocking probability of the hand-off calls increases for both guard band channels. When the hand-off call arrives to the cell and found out that channels are busy the hand-off calls will be blocked. Therefore, as the hand-off call arrives to a cell with higher average rate, they will be blocked with a higher probability.

On the other hand, as the number of guard bands in a cell increases, the more channels are reserved to service the hand-off calls. Therefore, the blocking probability decreases. For the Poisson model, the difference between green and red curves represents the difference of blocking probability for two different guard band channels. The blocking probability in case of 2 guard band channels is higher than the case where 8 guard band channels are considered.

For MMPP modeling, the results are identical to the Poisson model. When the average hand-off calls increases, the blocking probability also increases. From the same plot, we notice a considerable difference between the curves of both models under the same number of guard band channels. For instance, when the number of guard band channels is 2, the differences between the curves are in the range of 10^{-2}, which is not negligible for the blocking probability. This difference was caused by the bursty characteristics, which has been ignored in the Poisson model, however it has been considered with the MMPP model.

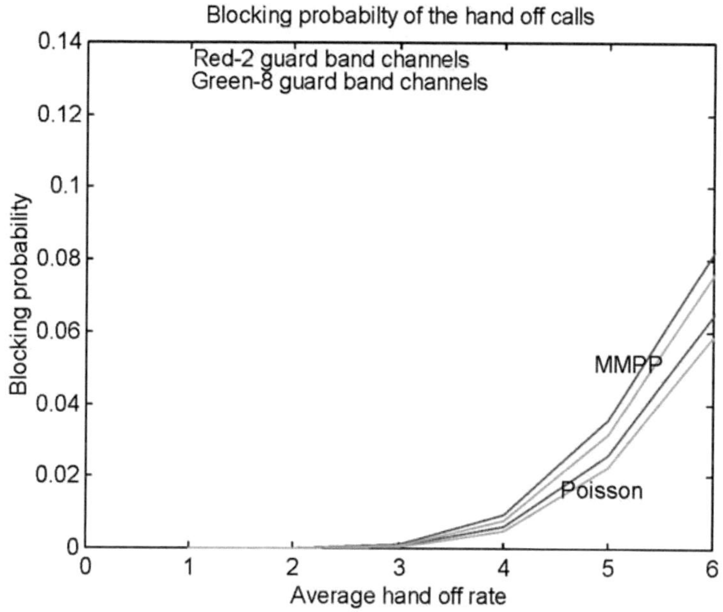

Fig.13: The average hand-off calls rate vs. the blocking probability of the hand-off calls for different guard band channels

4.6 The Number of Servers Versus the Blocking Probability of the Hand-off Calls

In figure 14, the following assumptions were made: The hand-off calls arrival rate in state one of the MMPP model are 5 calls per unit however in state two they are 1/5 calls per unit. The average hand-off calls arrival rate is 4.52, which is calculated using equation (1) in section 3.2. The new calls' arrival rate in both models is 1/5 calls per unit. The hand-off calls arrival rate in Poisson modeling is assumed to be equal to the average hand-off calls arrival rate in a two-state MMPP modeling. The plot in figure 14 shows the hand-off calls blocking probability vs. the number of channels for three guard band channels. The curves with 2 guard band channels are shown with a red color, and the green color is used to represent the curve with 6 guard band channels.

In a Poisson model, when the number of servers increases, the blocking probability decreases. As the number of channel increases, more channels are available for the hand-off calls to be serviced, therefore, the blocking probability decreases.

When the number of guard bands increases, the blocking probability decreases. The increase of the guard band channels will cause more reserved channels for the hand-off calls, therefore the hand-off calls have more chance to access a channel. These calls will be blocked with less probability.

The MMPP model shows the same behavior as the Poisson model for all the changes, but there are differences between their corresponding curves. The gaps between curves are in the range of 10^{-3}, which is not negligible as far as the blocking probability is concerned. This difference in result was caused by the bursty behavior of the hand-off calls.

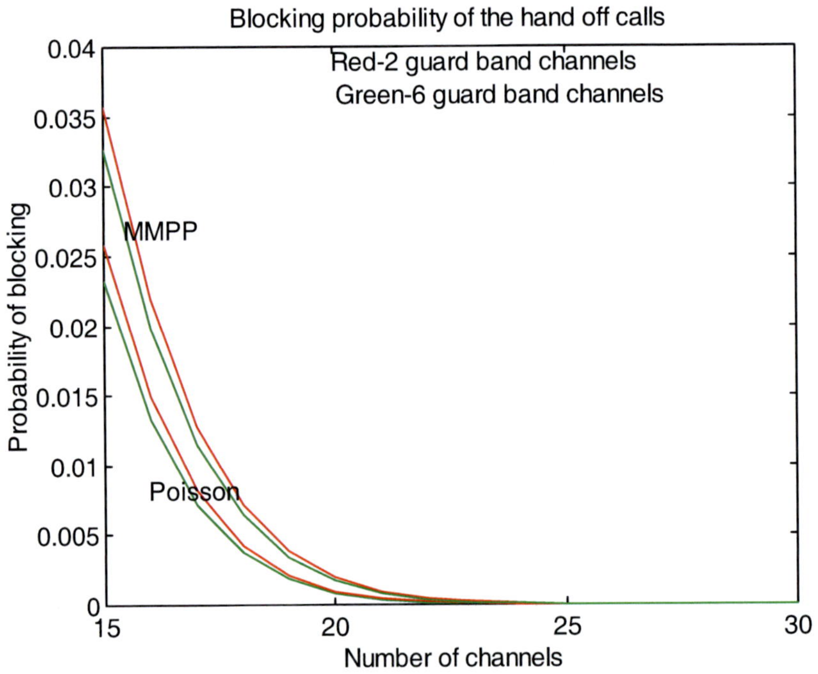

Fig.14: The number of servers vs. the blocking probability of the hand-off calls for different guard band channels

5 CONCLUSION

The following results are concluded from analytical analysis of the cellular system: for a specific number of available channels in a given cell, as the number of guard band channels increases, the cellular system will become unstable. When average arrival rate of the hand-off call increases, their blocking probability and the queue distribution of the new call increase. As the number of guard band channels increases, the new calls' queue distribution and their average queue length in the infinite queue increase, but the blocking probability of the hand-off calls decreases. When the number of available channels in a given cell increases, the average queue length of the new calls and the blocking probability of the hand-off calls decrease.

We have noticed that both models (Poisson and MMPP) are suitable in representing the arrival traffic to the cellular system and they made the analytical system analysis possible. Modeling the hand-off calls using MMPP model gives more accurate results compared to that given when considering the Poisson model for the same traffic. The difference between the results of the two models is caused by the specific behavior of the hand-off calls. The hand-off calls' arrival rate has a bursty characteristic, which was captured with the extra parameters of MMPP model.

As the result of this research, we can see that the performance metrics are sensitive to the arrival instances, therefore MMPP can be considered as a suitable model to present any bursty traffic.

6 FUTURE WORK

This work can be enhanced by considering the case where we have a finite buffer size for the new calls to join in case of unavailability of some free channel in a given cell. In this case, besides the performance metrics we were interested in during this research, we can look at the blocking probability of the new calls and study its behavior.

Another alternative consists of modeling both new calls and hand-off calls with MMPP and study the effect of the bursty behavior on the performance metrics.

REFERENCES

[1] John N. Daigle, and Nikhil Jain, "A Queuing System with Two Arrival Streams
And Reserved serves with Application to cellular Telephone." *INFOCOM'92*.

[2] M. F. Neuts, "Matrix-Geometric Solutions Stochastic Models." *John Hopkins University Press,* Baltimore, 1981.

[3] V. H. Mac Donald, "Advanced Mobile Phone Service: The Cellular Concept." *The Bell System Technical Journal,* Vol. 58, No. 1, January 1979.

[4] Theodore S. Rappaport, "Wireless Communications Principles and Practice." *Prentice Hall PTR,* 1996.

[5] J. M. Pitts, and J.A. Schormans, "Introduction to ATM Design and Performance with Application Analysis Software." *John Wiley,* New York, 1996.

[6] J. N. Daigle, R. Guerin, and N. Jain. Comments on "Queuing-Blocking System with Two Arrival Streams and Guard Channel." *Manuscript in Preparation,* 1992.

[7] O. Avellaneda, R. Pandya, and G. Brody, "Traffic Modeling of a Cellular Mobile Radio System." *In Proc. 11th Teletraffic Cong. (ITC 11),* Kyoto, Japan, Sept 1985, Vol. 1, pp. 2.4B-1-1 – 2.4B-4-7.

[8] W. Fischer, K. Meier-Hellstern, "The Markov-Modulated Poisson Process (MMPP) Cook book." *Performance Evaluation.* 18 (1992) 149-171.

[9] N. G. Hock, "Queuing Modeling Fundamentals." *John Wiley & Sons Ltd,* 1996.

[10] R. Guerin, "Queuing-Blocking System with Two Arrival Stream and Guard Channel." *IEEE Transactions on Communications.* 1988,36:2, 153-163

[11] J. N. Daigle, "Queuing Theory for Telecommunications." *Addison-Wesley,* Reading, 1992.

[12] S. M. Ross, "Introduction to Probability Models." *Academic Press,* New York, 1989.

[13] Nikhil Jain, "Eigenanalysis Solution for Quasi Birth and Death Process." *A Thesis Submitted to Virginia Polytechnic and State University,* Blacksburg VA, 1991.

[14] J. N. Daigle, R. Guerin, and N. Jain, Comments on "Queuing-Blocking System with Two Arrival Streams and Guard Channel." *Manuscript in preparation,* 1992.

[15] N. Akar and K. Sohraby, "A Generalized State-Space Formulation of M/G/1 type Markov Chain and Their Matrix Geometric Solutions." Submitted to *International Teletraffic Congress,* 1997.

[16] L. Kleinrock, "Queuing System Volume 1: Theory." *Wiley-Interscience Publication,* 1975.

[17] M. F. Nuets, "Matrix-geometric Solution in Stochastic Matrices of M/G/1 type and Their Applications." *Marcel Dekker, Inc.,* New York, 1989.

[18] H. Schellhaas, "On Ramaswami's Algorithm for the Computation of the Steady-State Vector in M/G/1 type." *Commun. Statist.-Stoc.* Models 6, 3 (1990), 541-550.

[19] Mohsen Guizani and Ammar Rayes, "Designing ATM Switching Networks" *McGraw-Hill*, 1999

Chapter 5

A FUZZY BANDWIDTH ALLOCATION CONTROLLER TO SUPPORT REAL-TIME TRAFFIC OVER WIRELESS NETWORKS

Shiann-Tsong Sheu and Meng-Hong Chen
Department of Electrical Engineering
Tamkang University
Tamsui, Taipei Hsien,
Taiwan 251, Republic of China
E-mail: stsheu@ee.tku.edu.tw

Abstract: - In this paper, we propose a fuzzy bandwidth allocation controller (FBAC) to support two types of services in wireless networks. The restricted time-bounded service (RTBS), such as voice and video service is sensitive to packet delay and the loose time-bounded service (LTBS), such as data transfer cares about the accuracy and can be serviced before a pre-assign due time. Based on the FBAC, five different request assignment strategies are introduced. Simulation results demonstrate that the fuzzy controller with some proposed strategies obtain high network performance, high fairness as well as an acceptable blocking probability.

Key-Words: - fuzzy bandwidth allocation controller (FBAC), Real-Time, Wireless Network

1 INTRODUCTION

In wireless network, many MAC protocols have been proposed about how to guarantee the quality of service (QoS) recently. Typically, slotted based protocols, like R-ALOHA, PRMA [1], ISMA [2], RS-ISMA [3], BRMA [4], reserve slots in the next cycle by pervious contention results. Since each of them does not consider the traffic characteristics, it is hard to exactly match the service requirements or obtain the maximum throughput. Adaptive protocols, such as PRMA/DA [5] and DSAMA [6], dynamic change the slots of different kinds of services to provide better QoS. However, too many kinds of services make the algorithms complexity and need much computing time. To reduce the protocol complexity, we simplify service types into two basic classes: (i) restricted time-bounded service (RTBS), such as voice and video services, (ii) loose time-bounded service (LTBS), such as data service. Each kind of services has its special traffic characteristic, bandwidth requirement and acceptable blocking probability. In order to fully utilizing the bandwidth in wireless network, a flexible bandwidth sharing strategy has to be applied. Although an aggressive bandwidth sharing scheme can provide a better management of network resource, due to the unpredictable fluctuations of traffic flow, the blocking probability of restricted time-bounded service may be violated.

To overcome the problem, we propose a fuzzy bandwidth allocation controller (FBAC) to deal with the bandwidth sharing problem in the wireless network. Based on the FBAC, five different request assignment strategies are proposed to dynamically select proper LTBS requests to serve. Among these strategies, two of them employ another fuzzy strategy switching controller (FSSC) to obtain the best results. The performance of the fuzzy controllers and strategies are evaluated and compared. The simulation results show that the proposed strategies can improve the network performance, bandwidth utilization, fairness. Most important, it has the ability to control the blocking probability of RTBS within an acceptable range as desired. Moreover, the proposed approach is very simple and can be implemented in hardware easily.

The rest of this paper is organized as follows. The system model is given in section II. In section III, we present the functional block diagram of the FBAC. In section IV, five different strategies based on the FBAC for assigning requests are discussed. In section V, the performance measurements, simulation models and results are reported. Finally, some conclusions are given in section VI.

2 SYSTEM MODEL

In wireless networks, channel is often divided into frames with fixed length and each frame is composed of *C* time slots (i.e., bandwidth is divided into *C* channels). In such time-division multiple access (TDMA), each time slot can be used by either RTBS or LTBS. According to the characteristics of two types of services, RTBS has higher priority than LTBS and the LTBS may starve when RTBS is overloaded. To prevent LTBS from starvation, a small amount of slots in frame, which is denoted as *R*, is particularly reserved for the LTBS and the remaining slots in frame can be allocated for either RTBS or LTBS. For simplicity, we denote the former and the latter as the *reserved* slots and the *sharable* slots, respectively. Therefore, the exact number of *sharable* slots is *C-R*. The channel's allocation of *sharable* and *reserved* slots within a frame in a wireless network is shown as Fig. 1.

As described above, when customer issues a request for LTBS, a particular parameter "due-time" is also specified at the same time. This identifies that it is valid only if it is delivered before this specified time. For example, assume a LTBS with transmission time *L* arrives at time *T*, the latest timing *T* ' of starting to service is *T+D-L* if it can be delayed no later than *T+D*.

3 THE FUZZY BANDWIDTH ALLOCATION CONTROLLER (FBAC)

In this paper, the fuzzy approach is adopted to determine how many sharable slots can be temporarily borrowed to serve LTBS without degrading the quality of RTBS. There are two input linguist parameters considered for the fuzzifier in FBAC: the RTBS blocking probability, which is denoted as *rbp,* and the channel free proportion (the ratio of the number of free sharable

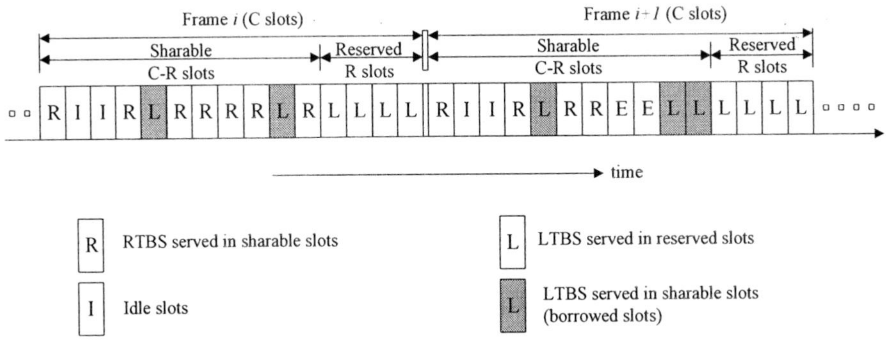

Fig. 1. The bandwidth allocation in a wireless network

slots to the total sharable slots), which is denoted as *cfp*. For input parameters, we define the corresponding fuzzy term sets: *T*(*rbp*) = {Safe, Normal, Dangerous} and *T*(*cfp*) = {Small, Medium, Large}. The selected membership functions for *T*(*rbp*) and *T*(*cfp*) are the shape of Gaussian-like function (see Fig. 2(a) and 2(b)). For each membership function, the peak position and the scaling factor are specified according to our knowledge about the system model. (The accepted blocking probability of RTBS is assumed to be 0.02.) The mathematics form of the Gaussian membership function is presented as follows:

$$\mu(x) = e^{-[\frac{x-m_i}{\sigma_i}]^2},$$

where m_i and the σ_i are the peak value and the scaling factor of the *i*th membership function, respectively. We also define the term set of the output borrowing ratio *br* as *T(br)* = {Small, Medium, Large}. According to the fuzzy set theory, the fuzzy rule base has $|T(rbp)| \times |T(cfp)| = 9$ inference rules (see Table 1). Obviously, if the *rbp* is safe and the *cfp* is high then the *br* is set to large. However, when the RTBS traffic load becomes heavy such that the *rbp* is dangerous, no matter how large the *cfp* is, the *br* is set to small.

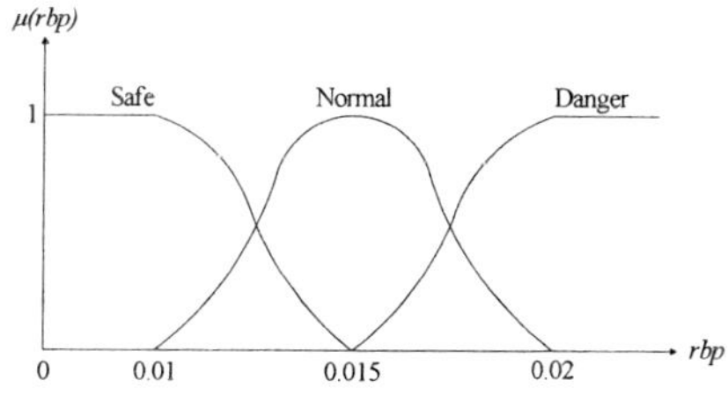

(a) T(rbp)

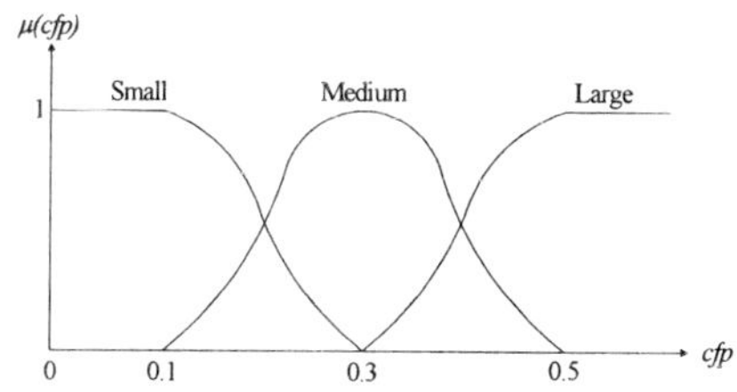

(b) T(cfp)

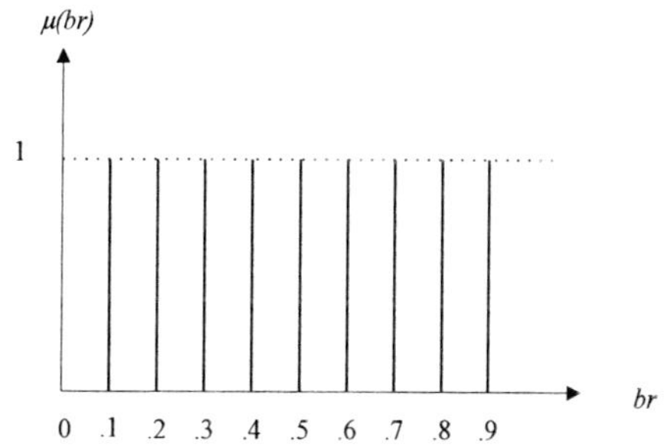

(c) T(br)

Fig. 2. The membership functions of the term sets

Table 1. The fuzzy control rules for borrowing channel.

Rule	rbp	cfp	br
1	Safe	Large	0.9
2	Safe	Medium	0.8
3	Safe	Small	0.7
4	Normal	Large	0.6
5	Normal	Medium	0.5
6	Normal	Small	0.4
7	Dangerous	Large	0.3
8	Dangerous	Medium	0.2
9	Dangerous	Small	0.1

In the inference engine, the max-min inference method [7] is used. For the ith rule, the corresponding membership values of these two input variables *rbp* and *cfp* are calculated by $\mu_i(rbp)$ and $\mu_i(cfp)$, respectively. The weight w_i used in defuzzifier is determined by the minimum value between $\mu_i(rbp)$ and $\mu_i(cfp)$. Considering the defuzzifer in FBAC, we employ the singleton method [8] as our defuzzification strategy to reduce the complexity of computation (see Fig. 2(c)). For each fuzzy rule, the method will convert the output membership function into a crisp output control value.

The borrowing ration is affected by the blocking probability of RTBS significantly. It is clear that if the *rbp* is dangerous (safe), the borrowing ratio should be small (large). In our case, the peak values for $T(br)$ = $\{0.1,0.2,0.3,0.4,0.5,0.6,0.7,0.8,0.9\}$. When the *rbp* states in normal state, the reason of setting the borrowing ratio to 0.6 is trying to obtain higher bandwidth utilization. The singleton defuzzification method calculates the crisp output value y_1 by the following equation:

$$y_1 = \frac{\sum_{i=1}^{n} w_i \times p_i}{\sum_{i=1}^{n} w_i},$$

where $n = |T(rbp)| \times |T(cfp)|$, p_i is the peak value of the ith output membership function of $T(br)$, and w_i is the weight of the ith control rule. Hence, the number of borrowing sharable slots (*BSS*) are easily obtained by the following equation:

$$BSS = (C\text{-}R) \times cfp \times y_1.$$

Although the number of borrowed slots are determined according to the present *rbp* and *cfp,* the way of choosing the LTBS to serve is still has a chance to affect the *rbp*. This is because that once a borrowed slot is occupied by a long-drawn LTBS, it will keep busy for a considerable time. During this period, all incoming RTBS calls may be blocked. Intuitively, a best strategy of choosing the shortest request to serve is able to degrade such influence. However, for the sake of fairness, the earliest arrival request should be served before other requests. Therefore, it is desirable to have an efficient request assignment strategy so that the influence caused by the LTBS could be minimized and the fairness could be achieved at the same time.

4 Five Request Assignment Strategies

In this section, five different strategies based on FBAC are proposed for determining which LTBS is chosen to use the borrowed slots. The five request assignment strategies are proposed as follows:

A. Early Request First Strategy (ERFS)

In ERFS, both the reserved and sharable channels, the order of serving requests is based on a simple first-come-first-serve concept. That is, an available channel is assigned to serve the request with the earliest arrival time. It is clear that based on this simple strategy, the fairness criterion is achieved completely. Nevertheless, the assigned request for the sharable channel may take a considerable time before it releases channel. Therefore, the blocking probability of real-time traffic may increase significantly. In order to reduce the blocking probability, the SRFS is proposed.

B. Short Request First Strategy (SRFS)

It is obvious that if the selected request for the sharable slot needs a shorter service time, the blocking probability of RTBS will be reduced. Besides, the blocking probability of the LTBS is also decreased because that the number of serviced calls increases. However, because the SRFS always selects the shortest services from buffer, as a result, almost unserviced requests have a longer service time and the time interval of sharable slots occupied by the LTBS requests will become longer and longer. Consequently, the number of blocked RTBS requests may become larger beyond expectation. The simple way to maintain the *rbp* is to reserve the sharable channels for RTBS instead of sharing by LTBS. However, this contradicts high channel utilization. It is not easy to achieve both at the same time. Another drawback of SRFS is that an earlier arrival request may be postponed by SRFS serving later incoming requests. It is an unfair scheduler. To overcome these shortcomings, an Adaptive SBLR is proposed.

C. Adaptive SBLR

The major concept of the adaptive SBLR is that the **S**hortest service in buffer is assigned for **B**orrowed channel and the **L**ongest service in buffer is assigned for **R**eserved channels. It is expected that the adaptive SBLR will obtain smaller blocking probability of RTBS than that of the SRFS obtained.

D. Fuzzy ERFS/SRFS Strategy (FESS)

A well-behavior assignment strategy should depend on the network status to apply an appropriate strategy. When the blocking probability of RTBS traffic becomes dangerous, the SRFS is applied to shorten the usage time period; otherwise, the ERFS is used to obtain fairness. Based on such concept, the borrowed channels may avoid from domination. Besides, when the RTBS require more bandwidth resources, the sharable slots can be allocated for them quickly. However, it is too hard to precisely classify the dangerous condition. Therefore, an extra fuzzy strategy switching controller (FSSC) is employed to dynamically determine whether the SLFS is more suitable for present network or not.

We note that, in FSSC, two input linguistic parameters *rbp* and *cfp* and the corresponding membership functions are the same as that of FBAC. For the FSSC, we define the fuzzy set of the output switching factor *sf* as *T(sf)* = {Safe, Dangerous} and the membership function of *T(sf)* is shown in Fig. 3. The switching factor is used to indicate the status of the network is safe or dangerous. When the switching factor exhibits danger, (i.e., the *rbp* is close to the bounded value 0.02), the network will switch to use the SRFS for controlling *rbp* and channel utilization. Otherwise, the network uses the ERFS to achieve fairness. The fuzzy control rules are shown in Table 2. The crisp output value y_2 of FSSC is also calculated by the singleton defuzzification method.

Table 2. The fuzzy control rules for switching strategy.

Rule	rbp	cfp	sf
1	Safe	Large	Safe
2	Safe	Medium	Safe
3	Safe	Small	Safe
4	Normal	Large	Safe
5	Normal	Medium	Safe
6	Normal	Small	Dangerous
7	Dangerous	Large	Dangerous
8	Dangerous	Medium	Dangerous
9	Dangerous	Small	Dangerous

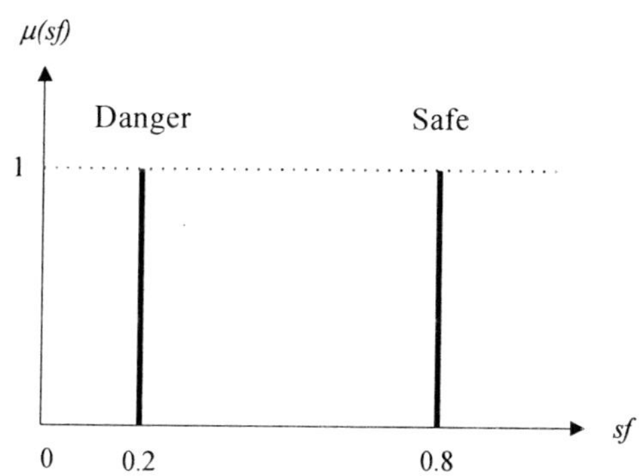

Fig. 3. The membership function of the term set T(sf).

In order to switch the request assignment strategy, a request assignment mechanism is used in FESS. According to the crisp output value y_1 of FBAC and the crisp output value y_2 of FSSC, the request assignment mechanism switches strategy dynamically and chooses an adequate number of the proper LTBS for serving. Therefore, the FSSC is much flexible to deal with the unpredictable traffic load. In addition, the system performance is also being improved at the same time.

E. Enhanced FESS (EFESS)

In FESS, only one strategy is applied on network at a time, the fairness and the blocking probability can not be improved simultaneously. To obtain a better result, we propose a much flexible strategy, named as Enhanced FESS (EFESS), to apply both strategies on the network at the same time. The concept of EFESS is described in detail as follows. When the capacity of sharable bandwidth is able to deal with the incoming RTBS traffic (which is determined by FSSC), the ERFS can be applied on the borrowed (sharable) slots to obtain fairness. Meanwhile, the SRFS is applied on the reserved slots to decrease blocking probability of LTBS. On the contrary, if the demand of RTBS excesses the entire capacity of sharable bandwidth, the scare borrowed slots should be utilized in a more efficient way. That is, we assign the borrowed slots to serve the requests with the shortest service time. Meantime, the reserved slots serve requests in a function of ERFS.

5 PERFORMANCE MEASUREMENT AND THE SIMULATION MODELS

5.1. Performance Measurement

Assume the simulation time for the simulation run is *T*. Based on the FBAC, the performances of five strategies are evaluated in terms of the following metrics:

- **RTBS Blocking Probability (RBP)** = the ratio of the number of blocked RTBS requests and the total arrival RTBS requests during a long period of simulation time *T*.
- **LTBS Blocking Probability (LBP)** = the ratio of the number of blocked LTBS requests and the total arrival LTBS requests during a long period of simulation time *T*.
- **Reserved Bandwidth Utilization (RBU)** = the ratio of the summation of busy time intervals of *reserved* bandwidth to serve requests and the total simulation time *T*.

- **Sharable Bandwidth Utilization (SBU)** = the ratio of the summation of busy time intervals of *sharable* bandwidth to serve requests and the total simulation time T.

Let **Unfairness Factor (UF)** denotes the number of LTBS requests that are passed the time limit and are dropped by the strategy chooses some later incoming requests to serve. Therefore, the **Network Fairness (NF)** is defined as

$$NF = \frac{SLR - UF}{SLR},$$

where SLR is the number of successfully served LTBS requests during the time T. A high performance network would require to have small BP, high BU and high NF.

5.2. Simulation Models

The proposed fuzzy bandwidth allocation controller and five different channel assignment strategies are investigated by simulations. The total number of available channels (time slots in frame) in wireless network is 60 (C=60). The numbers of *reserved* time slots are 5 (R=5). The acceptable blocking probability of RTBS is set to 0.02. For simplicity, we assume that a buffer with infinite volume is used for buffering all incoming LTBS requests. The request arrival rate of RTBS and LTBS are Poisson distributions with a mean $\lambda_i^{realtime}$ and $\lambda_i^{delayable}$ at the ith hour in one day, respectively (see Fig. 4). The request length for two kinds of services is an exponential distribution with a mean of L minutes. Therefore, the *Network Load* (NL) for the network is defined as

$$NL = \frac{\sum_{k=0}^{23} (\lambda_k^{realtime} + \lambda_k^{delayable}) \times L}{24 \times C}.$$

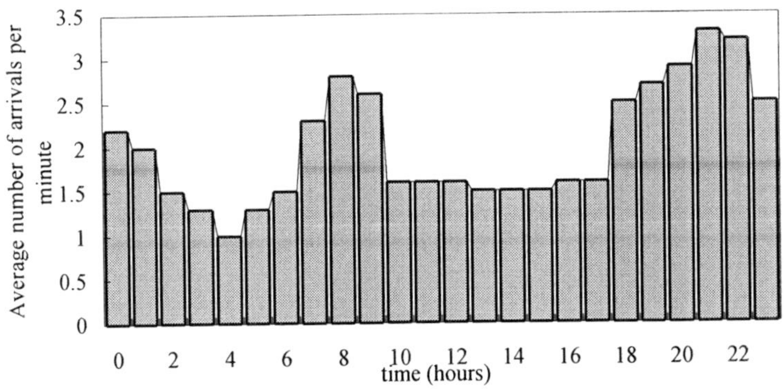

(a) RTBS Traffic

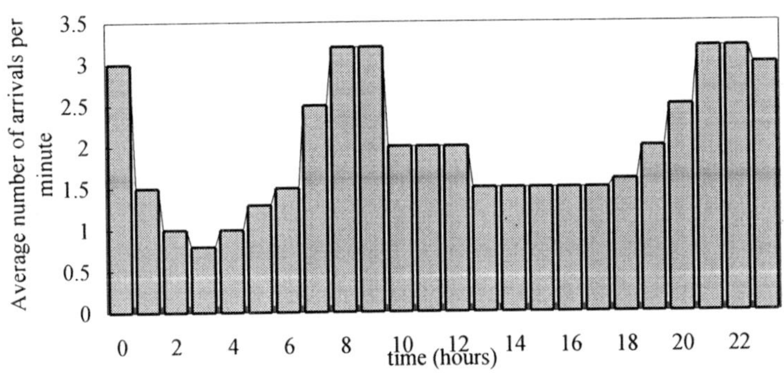

(b) LTBS traffic

Fig. 4. The average arrival rates for RTBS and LTBS traffic.

Since $\sum_{k=0}^{23}(\lambda_k^{realtime} + \lambda_k^{delayable})/24 = 4$ and $C = 60$, we obtain a simpler equation: $NL = (4 \times L) / 60$. Therefore, in these following simulation models, NL=1 if L=15. This implies that if mean service time L is larger than 15, the traffic load of the network becomes heavy or overloaded. In this paper, the simulation period is 30 days (i.e., T = 43200 minutes) and the parameter L is considered from 12 minutes (light loaded, NL = 0.733) monotonically increases to 20 minutes (heavy loaded, NL = 1.333) in a step of 1 minute and LTBS due-time D = 7 hours.

5.3. Simulation Results

Fig. 5 shows the RBPs and LBPs obtained by the ERFS, SRFS, Adaptive SBLR, FESS and EFESS under different mean service time of request when D = 7 hours. Fig. 5(a) illustrates the LBP obtained by the SRFS is smaller than that of other proposed strategies under different network load level. We can also see that the LBPs obtained by these strategies are almost identical when the network load is light ($L < 15$ *or* $NL < 1$). This is because that under this condition, nearly all the incoming LBTS can be served in spite of which strategy is applied. Intuitively, when the network load becomes heavy ($L \geq 15$ *or* $NL \geq 1$), a larger LBP would be obtained by proposed strategy. We note that the Adaptive SBLR produces a higher LBP when network loaded is heavy. When the network capacity can not deal with the incoming traffic load, the FBAC hardly borrows channels for LBTS. Thus, in SBLR strategy, only the requests with the longest service time have the opportunity to be served. This increases the LBP significantly. Fig. 5(b) illustrates the RBPs produced by the ERFS, SRFS, Adaptive SBLR, FESS and EFESS under different network load. Recall that the RBP in the network is required to be bounded beyond a specified value 0.02. In Fig. 5(b), the obtained RBPs by all the proposed strategies are obviously bounded beyond 0.02 even when $15 \leq L \leq 17$ (that is, $1.0 \leq NL \leq 1.13$). This implies that the proposed FBAC has the ability to control the RBP efficiently even when the network load is heavy. We also can find that the RBP of the SBLR is slightly better than that of other strategies.

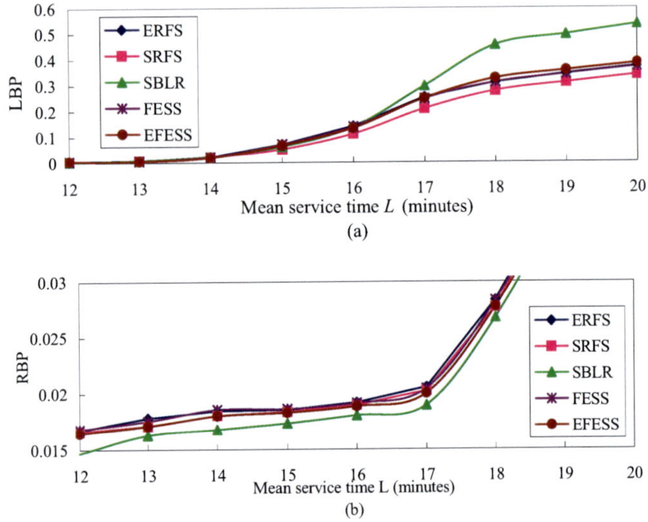

Fig. 5. Comparisons of LBP and RBP obtained by the ERFS, SRFS, SBLR, FESS and EFESS under different mean service time L.

Fig. 6 shows the results of reserved bandwidth utilization RBUs and sharable bandwidth utilization SBUs obtained by five strategies under different mean service time *L*. Since the reserved channels are dedicated for the LTBS, the channels will be fully utilized if the incoming LTBS traffic load excesses the reserved capacity has. In Fig. 6(a), we can see that the obtained RBU is almost 1.0 in spite of the mean service time of requests. This is because only five reserved channels are assigned in the simulation model. In Fig. 6(b), we can see that the obtained SBU under different mean service time of request is similar to each other. The peak value of the SBU of each strategy occurs when the network loaded is just saturated. If the incoming network load becomes much heavy, the FBAC will slow down the borrowing rate to prevent the blocking probability of RBTS from increasing seriously. However, the SBU is degraded. But, the percentage of SBUs of proposed strategies is still maintained above 75%.

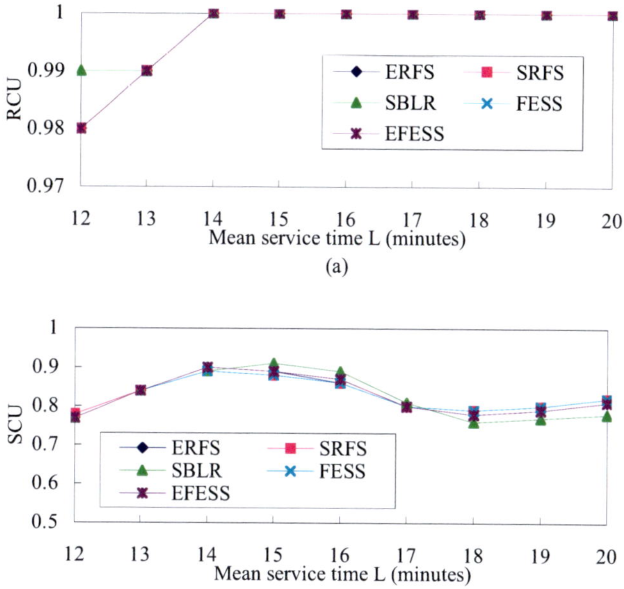

(a)

(b)

Fig. 6. Comparisons of RCU and SCU obtained by the ERFS, SRFS, SBLR, FESS and EFESS under different mean service time L.

The comparison of Network Fairness (NF) obtained by the five strategies under different mean service time L is shown in Fig. 7. Although the SBLR obtains the smallest real-time blocking probability, the drawback is that it performs quite unfair. The unfairness is resulted from the SBLR always selects the shortest and the longest request for the sharable and the reserved channels, respectively. Obviously, the ERFS is the fairest strategy. We note that the proposed FESS and EFESS dynamically switch the ERFS and SRFS strategies to gratify the network requirements. In the worse case ($L = 20$), both the values of network fairness of FESS and EFESS are slightly unfair ($NF \cong 95\%$) when comparing with the ERFS. This shows that the FESS and EFESS obtain an excellent result under different traffic demand.

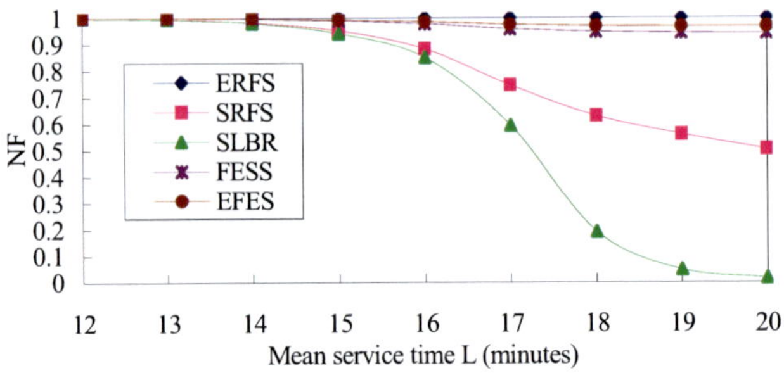

Fig. 7. Comparisons of NF obtained by the ERFS, SRFS, SBLR, FESS and EFESS under different mean service time L.

6 CONCLUSIONS

In this paper, we design a fuzzy bandwidth allocation controller (FBAC) in the wireless network to provide QoS. Base on the FBAC, we also propose five strategies to select the being served LTBS requests. The simulation results demonstrate that the proposed strategies can guarantee low blocking probability, high bandwidth utilization, and fairness.

REFERENCES

[1] D.J.Goodman, R.A.Valenzuela, K.T.Gayliard, and B.Ramamurthi, "Packet Reservation Multiple Access for Local Communications", *IEEE Trans. on Communications*, vol. 37, no. 8, August 1989, pp.885-890.

[2] G. Wu, K. Mukumoto and A. Fukuda, "An Integrated Voice and Data Transmission System with Idle Signal Multiple Access –Static Analysis", *IEICE Trans. Comm.*, vol. E76-B, no. 9, pp.1186-1192, Sept.1993.

[3] Gang Wu, Yoshihiro Hase, Kazumasa Taira, and Ken Iwasaki, "A Wireless ATM Oriented MAC Protocol for High-Speed Wireless Lan", *Proceedings of the IEEE PIMRC'97*.

[4] Z. Zhang, I. Habib and T. Saadawi, "A Bandwidth Reservation Multiple Access Protocol for Wireless ATM Local Networks", *International*

Journal of Wireless Information Networks, Plenum Publishing, Vol.4, No.3, 1997.

[5] J.G.Kim and I.Widjaja, "PRMA/DA: A New Media Access Control Protocol for Wireless ATM", *Proceedings of the IEEE ICC'96*, Dallas, TX, June 1996 pp.1-19.

[6] Xiaowen Wu, and Shiqi Wu, Hairong Sun, Lemin Li, "Dynamic Slot Allocation Multiple Access Protocol for Wireless ATM Networks", *Proceedings of the IEEE ICC'97.*

[7] H. J. Zimmermann, *Fuzzy Set Theory and its Application*, 2nd, revised edition, Kluwer Academic Publishers, pp. 11-17, 1991.

[8] R. R. Yager, L. A. Zadeh, *An Introduction to Fuzzy Logic Application in Intelligent System*, Kluwer Academic Publishers, 1992.

Chapter 6

A LEARNING FUZZY CONNECTION ADMISSION CONTROL FOR WIRELESS MULTIMEDIA NETWORKS

Mohsen Guizani and Wesam Alanqar
Department of Computer Science
University of West Florida
UNITED STATE OF AMERICA
mguizani@cs.uwf.edu

Abstract: The great improvement in wireless networks and the huge demand of multimedia internetworking have pushed for a more effort in combining both technologies for best usage of the Internet. In order to provide the agreed upon Quality of Service (QoS) at the time of connection acceptance throughout the connection duration, a better and more enhanced Connection Admission Control (CAC) has to be designed. The problem with regular CAC algorithm is the fixed allocation of bandwidth in adjacent cells to guarantee the QoS for mobile connections. Better dynamic bandwidth allocation has to be used for mobile users. In our study, learning fuzzy logic is used to monitor and dynamically adjust the reserved bandwidth in adjacent cells where the connection exists. Two types of traffic classes were used to study the learning fuzzy CAC performance, class I is QoS dependent, Class II is QoS independent. It was noticed through the simulation results that learning fuzzy CAC provides less total number of blocked new connections and less total number of dropped mobile connections. This was achieved through the dynamic allocation of reserved bandwidth to cope with the current network conditions.

Key Words: - Fuzzy systems, multimedia, wireless networks, Quality of Service (QoS), Connection Admission Control (CAC).

1 INTRODUCTION

1.1 QoS Traffic Control Mechanism

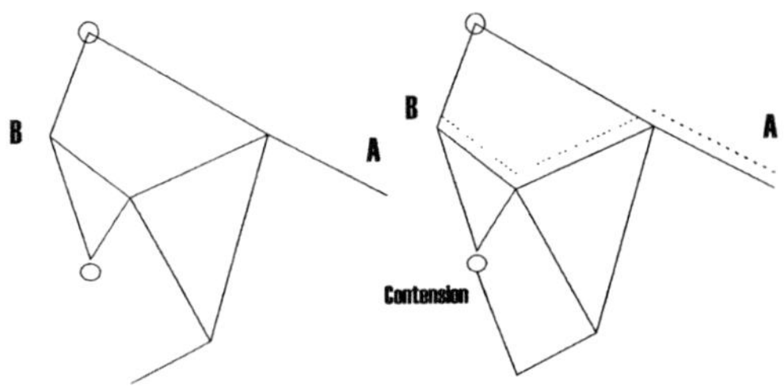

Fig. 1. An example of connection control to avoid congestion.

To ensure high level of performance for an ATM network, traffic control capabilities have been introduced. These capabilities can cope with various service classes and errors occurring any time in the network. Traffic control provides the tools for the network user to make sure that the available cell flows matches the specifications in the traffic contract. Traffic control includes: call admission control, usage parameter control, priority control, and network resource management [1,2].

1.1.1 Call Admission Control (CAC)

Call Admission Control (CAC) is a set of actions taken by the network during the calling set-up phase. This is needed in order to establish either VPC or VCC for the desired connection. The requested connection will not be established if the available network resources are not enough for a new call. The new connection also has to guarantee its required QoS. Meanwhile, it should not affect the QoS of other connections.

There are two types of classes that the connection admission control should consider:

1. Set of parameters that describe the source traffic characteristics such as Peak Cell Rate (PCR), Average Cell Rate (ACR), burstiness, and peak duration.
2. A set of parameters that describe the required QoS such as cell transfer delay, jitter, and cell loss ratio.

1.2 Congestion Control Algorithms

Congestion can be defined as the state of network elements in which the network can not provide the agreed upon QoS to already existing connections and to new requested ones. In other words, congestion occurs when the offered load from the user to the network exceeds the available network resources to guarantee the required QoS. Some of the network resources that can cause congestion are switch ports, buffers, ATM adaptation layer processors, and CAC processors. Two types of congestion may occur in an ATM network. Long-term congestion which is caused by more traffic load than the network can handle. Short-term congestion is caused by burstiness in the traffic. Techniques used for congestion control include admission control, resource reservation, and rate-based congestion control [3,4].

1.2.1 Admission Control

In this scheme, once the network detects congestion, no new virtual connection circuits are accepted. This technique is widely used and easy to implement. For example, in the telephone network, once congestion occurs, no more new phone calls can be made in this telephone network. Admission control may also allow new connection if the network can find a route that avoids the network congestion ports and has the demanded QoS. If no such route can be found, the connection will fail and be rejected.

Traffic management in ATM and multimedia networks comprises several functions which act at all levels from cell level to connection level in order to make sure that the QoS requirements are met for all connections carried by the network. Traffic management becomes especially demanding if traffic ranging from video conferencing over interactive applications (e.g. WWW) to bulk data transfer has to be handled by the network with minimal interference between different traffic streams.

Connection admission control (CAC) has to determine whether a new connection setup request can be accepted or should be rejected. This decision is based on the constraint to meet the negotiated QoS requirements of all existing connections as well as of the new connection. In addition to this basic function of CAC, there is the secondary goal to maximize the system utilization by allowing for a statistical multiplexing gain, i.e., an efficient CAC method should accept as many connections as possible without violating any QoS guarantees. However, there are some constraints that have to be considered when designing a CAC algorithm [5]:

- CAC is performed as part of the connection setup process and generally runs as a software module on a switch control processor. As setup times and control processor load must be limited, time consuming and complex CAC algorithms are not appropriate for real systems.
- Often assumptions are made about the traffic carried in the network. Finally, however, CAC can only rely on the traffic parameters negotiated in the traffic contract.
- CAC has to be applicable to a concrete switch architecture

Next, we introduce some fuzzy logic concepts [6,7] needed for the design of the new algorithm.

2 Fuzzy Logic Concepts

Fuzzy Logic is basically a multivalued logic that allows intermediate values to be defined between conventional evaluations like *yes/no, true/false, black/white,* etc. Notions like *rather warm* or *pretty cold* can be formulated mathematically and processed by computers. In this way, an attempt is made to apply a more human-like way of thinking in the programming of computers. A fuzzy set membership function is shown in Fig. 2.

2.1 Fuzzy Set Theory

Fig. 2. Fuzzy sets membership functions.

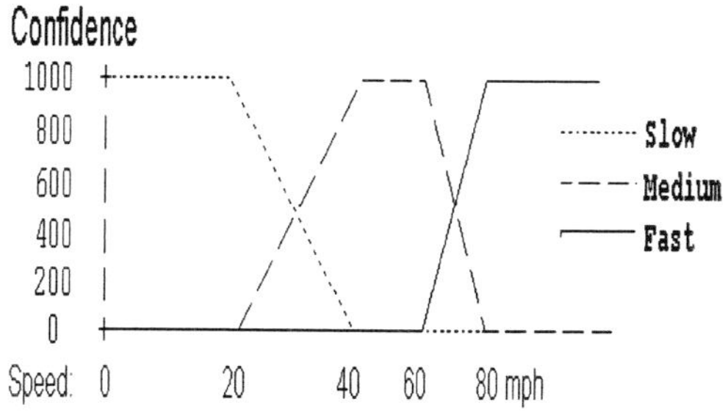

2.2 Fuzzy Control in Detail

2.2.1 Membership functions

Fuzzy controllers are very simple conceptually. They consist of an input stage, a processing stage, and an output stage. The input stage maps sensor or other inputs (such as switches, thumbwheels, and so on) to the appropriate membership functions and truth values. The processing stage invokes each appropriate rule and generates a result for each. Then, it combines the results of the rules. Finally, the output stage converts the combined result back into a specific control output value. The most common shape of membership functions is triangular, although trapezoidal and bell curves are also used. But, the shape is generally less important than the number of curves and placement. From 3 to 7 curves are generally appropriate to cover the required range of an input value (the "universe of discourse" in fuzzy jargon). Fig. 2 is an example of input membership functions to the fuzzy controller.

2.2.2 Rule Base

The processing stage is based on a collection of logic rules in the form of IF-THEN statements, where the IF part is called the "antecedent" and the THEN part is called the "consequent". Typical fuzzy control systems have dozens of rules.

Consider a rule for a thermostat:

IF (temperature is "cold") THEN (heater is "high")

This rule uses the truth value of the “temperature” input (some truth value of “cold”) to generate a result in the fuzzy set for the “heater” output (some value of “high”). This result is used with the results of other rules to finally generate the crisp composite output. Obviously, the greater the truth value of “cold”, the higher the truth value of “high” (though this does not necessarily mean that the output itself will be set to “high”).

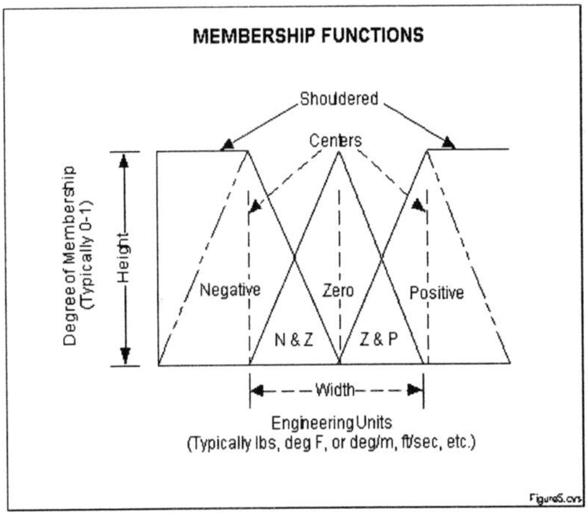

Fig. 3. Input membership functions to a fuzzy controller.

2.2.3 Fuzzification

In some cases, the membership functions can be modified by the so-called "hedges" that are equivalent to adjectives. Common hedges include "about", "near", "close to", "approximately", "very", "slightly", "too", "extremely", and "somewhat". These operations may have precise definitions, though the definitions can vary considerably; "very", for one example, squares membership functions -- since the membership values are always less than 1, this narrows the membership function. "Extremely" cubes the values to give greater narrowing; while "somewhat" broadens the function by taking the square root.

In practice, the fuzzy rule sets usually have several antecedents that are combined using fuzzy operators, such as AND, OR, and NOT (though again the definitions tend to vary). AND (in one popular definition) simply uses the minimum weight of all the antecedents, while OR uses the maximum value. There is also a NOT operator that subtracts a membership function from 1 to give the complementary function.

2.2.4 Defuzzification

There are several different ways to define the result of a rule, but one of the most common and simplest is the so-called "max-min" inference method, in which the output membership function is given the truth value generated by the premise. Rules can be solved in parallel in hardware or sequentially in software. The results of all the rules that have fired are "defuzzified" to a crisp value by one of several methods. There are dozens in theory, each with various advantages and drawbacks. The "centroid" method is very popular, in which the "center of mass" of the result provides the crisp value. Another approach is the "height" method, which takes the value of the biggest contributor. The centroid method favors the rule with the output of greatest area, while the height method obviously favors the rule with the greatest output value. Each rule provides a result as a truth value of a particular membership function for the output variable. In centroid defuzzification, the values are ORed (that is, the maximum value is used and values are not added) and the results are then combined using a centroid calculation. Fig. 4 shows how defuzzification is used to find a crisp output value from the fuzzy controller.

2.2.5 Fuzzy Model Reference Learning Control (FMRLC)

The FMRLC system consists of four main parts: the system to be controlled, the fuzzy controller, the reference model, and the learning mechanism. The learning mechanism tunes the rule base of the fuzzy controller so that the closed loop system behaves like the reference model. The learning mechanism consists of two parts; a fuzzy inverse model and a knowledge base modifier. The fuzzy methods of tuning fuzzy controllers, current accumulated industrial experience proved the ease with which those limitations can be overcome.

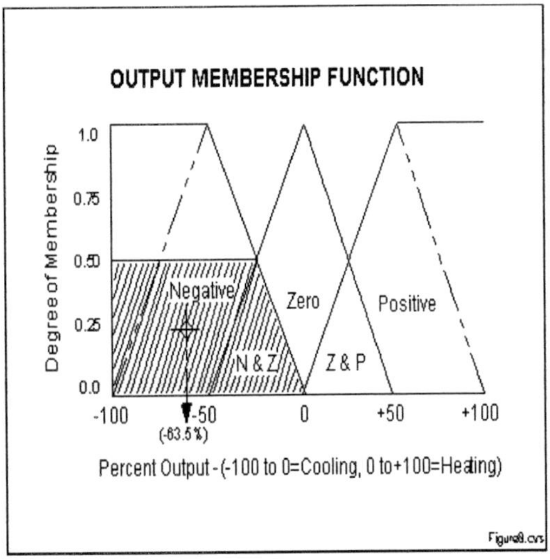

Fig. 4. Defuzzification operation in the output membership functions

While fuzzy control can provide effective practical solutions to complex industrial problems, as an alternative to conventional control methods, there are several drawbacks that may limit its use for some problems. One main drawback is the inability of the fuzzy controller, designed for the nominal system parameters, to perform adequately if significant and unpredictable plant variations occur.

Introducing the capability of a progressively learning mechanism into the system, along with basic ideas of fuzzy sets and control theory, will improve the performance of the overall controlled system. The fuzzy model reference learning controller (FMRLC) utilizes a learning mechanism which observes the system outputs and adjust accordingly the rules in the direct fuzzy controller such that the overall system performs satisfactory.

The inverse model performs the function of mapping $y_e(KT)$ (representing the deviation from the desired behavior), to changes in the process inputs p(kT) that are necessary to force $y_e(kT)$ to zero. The knowledge base modifier performs the function of modifying the fuzzy controller's rule base to affect the needed changes in the process inputs. A fuzzy system is used to map $y_e(kT)$ and possibly functions of $y_e(kT)$, such as $y_c(kT) =(1/T) (y_e(kT)-y_e(kT-T)$, to perform the necessary changes in the process inputs p(kT). This map is called "fuzzy inverse model" since information about the system inverse dynamics is used in its specification. Not that similar to the fuzzy controller, the fuzzy inverse model contains scaling factors g_{ye}, g_{yc}, and g_p. Given g_{ye}, g_{yc}, and g_p as inputs to the fuzzy inverse model, the rule-base for the fuzzy inverse model contains rules of the form:

$$\textbf{\textit{IF}}\ y_e\ \textit{is}\ Y_e^j\ \textbf{\textit{and}}\ y_c\ \textit{is}\ Y_c^I\ \textbf{\textit{THEN}}\ p\ \textit{is}\ P^m$$

Where Y_e^j and Y_c^I denotes linguistic values associated with the linguistic variables y_e and y_c respectively and P^m denotes the linguistic value associated with the mth output fuzzy set.

The fuzzy controller's rule base is modified by the one-step back approach. Here the new control action, *u(kT)*, would be the previously applied control action, *u(kT-T)*, plus *p(kT)*. Knowledge base modification is performed by shifting the centers C_m of the membership functions of *U*, i.e. the consequent linguistic value associated with *u*, the linguistic variable associated with the fuzzy controller's output *u(kT)*. This is carried out in two steps:

1. Find all the rules in the fuzzy controller whose premise certainty is $u_i[(kT-T),c(kT-T)] >0$ "call this active set of rules at kT-T".

2. For all the rules in the active set, use $c_m(kT)=c_m(kT-T)+p(kT)$ where $c_m(kT)$ is the center if the mth membership function at time kT. The selection of the scaling factors g_1 and g_2 are carried out using on of the following two approaches:

First approach:

- Select g_1 so that $y_e(kT)$ will not saturate the input membership functions certainty (neat the end points).
- Choose $g_p=g_u$ and let $g_2=0$
- Apply a step reference input $r(kT)$ and observe the system and reference model response.
- If there exist unacceptable oscillation in the system response, then increase g_2.
- If the plant output is unable to keep up with the reference model response, then decrease g_2.

Second approach:

Begin with $g_p=0$, i.e., no adaptation, and simulate the system.
Choose g_1 and g_2 so that there is no saturation on the input universe of discourse.
Increase g_p slightly and observe the system response. You may need to tune g_1 and g_2.
Continue to increase g_p. You should get g_p large enough for quick adaptation rate, but without getting into undesirable oscillations and instability.

2.2.6 A learning fuzzy connection admission control

The problem with conventional CAC is the non-adaptive and non-dynamic bandwidth allocation for new and existing connections. In other words, once the bandwidth has been allocated to a new connection or mobile connection, this bandwidth remains allocated regardless whether the connection needs all the allocated bandwidth or not. Fixed or non-dynamic allocation of bandwidth decreases the bandwidth utilization of the network.

Fuzzy logic can be used to monitor the allocated bandwidth utilization. Once CAC reserves bandwidth for new or existing connections, fuzzy logic rules can be used to adjust the reserved bandwidth size. If the allocated bandwidth is not fully utilized, fuzzy logic algorithm will decrease the allocated bandwidth and hence increase bandwidth utilization. This is expected to increase the total number of new connections the CAC can permit to get into the network. Increasing the total number of new connections means that the total number of blocked connections will be dropped that results into lower blocking ratio. One of the fuzzy rules that can be used to adjust the allocated bandwidth is the following:

IF *the allocated bandwidth utilization is medium and the increase in bandwidth utilization is negative*
THEN *the change in the allocated bandwidth is negative small.*

This fuzzy rule transfers the network engineer experience to control and adjust the allocated bandwidth. The first part of the "IF" statement means that the allocated bandwidth utilization is in a medium range defined by the membership functions. So, the bandwidth utilization is not as high as recommended by network engineer. The second part of the "IF" statement means that the bandwidth utilization is decrementing with time. This is of course not desired by network monitoring software or network engineers. So, a logical solution to this undesired situation is to reduce the allocated bandwidth for the existing connection, without affecting QoS of existing connections, and hence increase the bandwidth utilization. This is what is meant by the "THEN" statement. Fig. 5 shows in a general structure for a fuzzy CAC system.

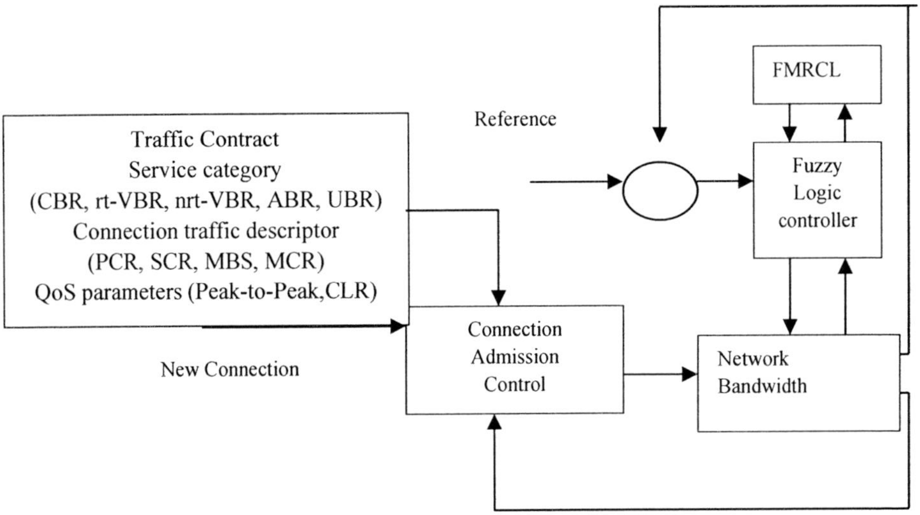

Fig. 5. A learning fuzzy CAC system

In this figure, the CAC algorithm is responsible for examining the requests of the new connections coming into the network. When a new connection arrives, CAC will know which type of traffic the new connection is, and then CAC will use the traffic contract to know all the required traffic parameters needed by that type of traffic. Examples of such parameters are the Cell Loss Ratio (CLR), Peak Cell Rate (PCR), maximum Burst Size (MCR). CAC will always have a continuous feedback from the network regarding the available bandwidth. If the required bandwidth from the new connection is less than or equal to the available bandwidth, the new connection is accepted and the required bandwidth is allocated. If the required bandwidth of the new connection is less than the available bandwidth, the new connection is rejected and blocked from entering the network.

As we can see, CAC will guarantee that the existing connections' bandwidth will not be affected by the bandwidth requests of new connections. Also, CAC will check the entire existing connections behavior. If any of the existing connections exceeds the bandwidth that has been agreed on when the connection was allowed to enter the bandwidth, such existing connection will be dropped by CAC to guarantee QoS for the rest of existing connections.

The problem with CAC is the non-adaptive fixed allocation of bandwidth. Once the new connection is accepted and the required bandwidth is reserved in the network, this bandwidth will remain unchanged during the duration of the connection. As we know, for Variable Bit Rate (VBR) traffic, the required bandwidth changes within the minimum and maximum bandwidth of that traffic class. This means that at any time the real required bandwidth by an existing connection could be much less than the reserved bandwidth. This means that there will be wasted bandwidth and the network bandwidth utilization will be low. This has a big impact on the efficiency of CAC. CAC will use the available bandwidth feedback from the network to judge weather the new connection should be accepted or rejected. If the available bandwidth keeps dropping down due to the unused bandwidth, the total number of allowed connections will also drop.

Fuzzy logic can be used to monitor and adjust the reserved bandwidth for existing connections. Fuzzy logic rule base will contain an expert network engineer experience in monitoring and adjusting the reserved bandwidth. As we can see from Fig. 5, the fuzzy logic controller will have a continuous feedback from the network of the reserved bandwidth utilization. Fuzzy logic will use the desired reference point for network utilization with the feedback of the current network bandwidth utilization to feed it to its expert rule base. The fuzzy logic controller will then generate a crisp output value to adjust the current reserved bandwidth to cope with the dynamic changes in the network. This of course will affect the bandwidth utilization value and increase it to meet the desired reference point of the fuzzy logic controller.

The problem of conventional fuzzy logic controller is its fixed rule base. This means that once the rule base has been designed for expected network conditions and parameters, this rule base can not be changed later. The network might get into severe network conditions or parameters that were not taken into consideration while designing the fixed rule base of the fuzzy logic controller.

Here comes the significant improvement of the learning fuzzy model reference learning control (FMRLC). The proposed learning mechanism will have a continuous feedback from the network regarding the current network conditions and parameters. If the network goes under unexpected situations such as a severe bandwidth drop or a sudden high need for a bandwidth, the learning mechanism will be able to adjust the fixed rule base in the fuzzy logic controller to meet the new conditions and demand. In Fig. 5, there is a continuous bi-directional connection between the learning mechanism and the fuzzy logic controller to keep the later database dynamic enough to cope with any network unexpected situations.

FMRLC has its own built-in memory, this has a great improvement over the conventional fuzzy logic controller. The improvement comes since FMRLC will store the last frequent network conditions the network went through. This will help the learning mechanism to fire the correct rules that satisfy the current conditions. In addition, it will minimize the computation needed to check which rules are needed for such conditions. As a result, a faster action will be taken and hence a better dynamic allocation of bandwidth. This of course will increase the total allowed number of new connections that means better CAC performance.

3 Problem Definition

During the last few years, there was a huge impact and demand on the usage of multimedia traffic through the Internet. Due to the fast improvements in wireless technology and the developments in data transfer through wireless devices, the trend in internetworking shifted to try to provide the multimedia data transfer though wireless phones. One of the interesting trends nowadays is Sun Microsystems research to move the bottleneck of bandwidth and computation speed from the wireless laptops to the server side. This will allow a much better performance for wireless internetworking, especially while transferring multimedia data.

The telecommunication Integrated on Demand Network (ION) uses ATM technology in providing one service for all type of Internet traffic like voice, data, and video. Sprint might face some problems in deploying its ION services in local areas where local telephone companies provide the local telephone services. The current trend of Sprint is to use its wireless Sprint PCS services to provide such high-speed multimedia data transfer to provide the ION services and bypass local telephone companies regulations. Such trend pushes for the need of more research in the area of providing efficient wireless multimedia through the Internet.

The problem with using wireless technology in multimedia traffic has some difficulties that have to be considered for best data transfer performance.

Multimedia data requires very constrained QoS parameters. For example, when we use the Internet to carry video conferencing, all users of such service have to experience very small delay and change in delay of the arrival of images from the source. If we want to use wireless networks to transfer such constrained QoS data, the problem gets more complicated. The main reason for this is the user's mobility through the physical region where the service exists.

Since the trend now is to expand the usage of the Internet and increase the number of users through wireless networks, a great challenge is to try to improve multimedia data traffic management and flow and congestion control algorithm. The focus of such research is towards improving the existing flow and congestion control algorithms to avoid and minimize possible bottlenecks due to the usage of wireless networks.

In particular, we want to improve CAC algorithm to provide better (QoS) satisfaction for wireless multimedia data traffic. Standard CAC algorithm allocates fixed bandwidth to new connections coming into the network. For current real time multimedia data, the actual bandwidth needed at any instant might be much less that what CAC already allocated for such connection at the time of arrival. This inefficient allocation of bandwidth will drop down the network bandwidth utilization and will decrease the total allowed number of new and mobile connections to the network. Another issue that has to be considered while using wireless networks for multimedia traffic is the factor of user mobility through the regional area. This will add more effort to design an efficient and dynamic CAC algorithm that will monitor the network bandwidth allocation and utilization around the clock.

This new proposed dynamic CAC uses fuzzy logic to improve the dynamic allocation of bandwidth and improves the bandwidth utilization. The new admission control scheme is proposed to provide high degrees of QoS guarantees for multimedia traffic carried in microcellular, high-speed wireless networks. The proposed scheme combines both admission control and fuzzy logic concept for bandwidth reservation to guarantee QoS requirements.

The new proposed learning fuzzy CAC algorithm is distributed. It considers both local information (e.g., the amount of unused bandwidth in the cell where the user currently resides) and remote information (e.g., the amount of available bandwidth in neighboring cells) to determine whether to accept or reject a connection. Since a mobile user is free to move anywhere, an admission control algorithm that depends on local information in the current cell, where the user resides, cannot guarantee QoS requirements of a connection throughout its lifetime. The proposed algorithm thus uses both local and remote bandwidth conditions, and allocates bandwidth in the cell where a connection request originates and reserves bandwidth in the neighboring cells.

When a user moves to a new cell necessitating a call handoff, the reserved bandwidth in the cell that the user is moving into is used to support the handoff connection. In addition, every time a user moves to a new cell, bandwidth is reserved in the new neighboring cells, and the reserved bandwidth in the cells, which are no longer neighboring to the new cell, is released. Further, the proposed scheme distinguishes real-time traffic and non-real-time traffic, and reduces the bandwidth assigned to non-real-time connections to provide higher QoS to real-time connections if necessary.

4 Proposed Algorithm

The proposed scheme has the following features:

1. It combines connection admission control with a bandwidth reservation and monitoring mechanism based on a learning adaptive fuzzy logic algorithm.
2. It uses both local information and remote information to determine weather to accept or reject a new or mobile connection.
3. It is adaptive, in the sense that it adjusts the amount of reserved bandwidth based on the current network conditions.
4. It distinguishes between new and mobile connections. It gives higher priority to mobile connections to provide higher connectivity to users who are already accepted in the network.
5. It reduces the bandwidth assigned to non-real time connections to provide higher QoS to real-time connections.
6. It is distributed, i.e., it is performed at each base station in a distributed manner.

In the proposed wireless system, two types of traffic are assumed. Class I is real-time traffic that includes video and voice traffic from users equipped with an adjustable rate codec. In case of congestion, such users can gracefully adjust the coding rate such that the quality of video/audio received at destination(s) is still acceptable. However, if the coding rate is reduced below some threshold, the quality of received video/audio becomes unacceptable. Class II traffic includes non-real-time data at a network node (such as a base station) or at a user station and transmit them at a slower rate. For class II traffic, it is assumed that there is no minimum required bandwidth since it can tolerate relatively large delays.

In the proposed scheme, it is assumed that when a mobile user requests a new connection or roams into a new cell, it provides the following information:

7. Class of traffic (Class I or Class II).
8. Desired amount of bandwidth of the connection.
9. Minimum required amount of the connection (only for class I traffic).

The minimum required bandwidth for a class I connection is the minimum amount of bandwidth that a source requires to maintain acceptable quality (e.g., such as the minimum coding rate at the source).

For a new connection, the scheme attempts to allocate the desired amount of bandwidth in the cell where the new connection is generated. If the desired amount of bandwidth is not available, the new connection is rejected. If the desired amount of bandwidth is available, and the new connection is of class II, then it is accepted, and the desired amount of bandwidth is allocated. If the new connection is of class I, the proposed scheme allocates the desired amount of bandwidth in the cell where the new connection originates, and reserves bandwidth in all the neighboring cells in anticipation of handoffs.

For a class I handoff connection, the proposed scheme works as follows: If the amount of available bandwidth in the cell that the handoff connection is moving into is less than the minimum required bandwidth of the connection, the handoff connection is dropped. Otherwise (i.e., if the amount of available bandwidth is greater than or equal to the minimum required bandwidth of the connection), the lesser of the available bandwidth in the cell and the connection's desired amount of bandwidth is allocated to the handoff connection.

At the same time, bandwidth reserved in the new neighboring cells is released. If bandwidth reservation succeeds in all new neighboring cells, the handoff connection is dropped. As mentioned earlier, class I connections, such as those from users with an adjustable rate video/voice codec, can adjust the coding rate based on the bandwidth availability and, if necessary, reduce the coding rate down to its minimum level, still maintaining an acceptable quality at the destination. For class II handoff connection, it is accepted as long as there is some bandwidth available in the cell that the connection is moving into. A class II handoff connection is dropped only when there is no bandwidth available in the cell that the connection is moving into. Bandwidth reduction in a class II handoff connection results in a slower transmission rate and, thus a longer transmission delay. However, class II is for non-real-time traffic and the impact of increased delay should not be significant.

Note that in the proposed scheme, when the available bandwidth in the cell is less than the desired amount of bandwidth, a new connection is always rejected. Whereas, a handoff connection may be accepted if the minimum required bandwidth can be provided. In other words, the proposed scheme provides priority to a user that is already in the network (i.e., a handoff connection) over a new user (i.e., a new connection).

As described above, bandwidths are reserved at connection setup time and during handoff for class I connections. Each time a mobile user moves to a new cell, bandwidth is reserved in the new neighboring cells and the reserved bandwidth in the cells which no longer neighbor the new cell is released. Fig. 6 is an example of a cellular system

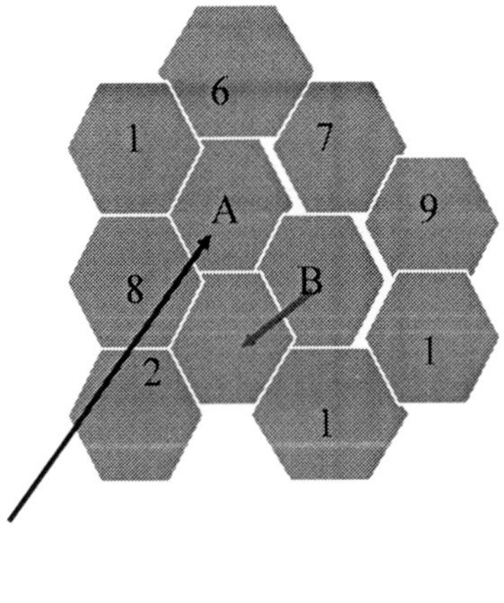

Fig. 6. A sample of a cellular system

In Fig. 6, assume that a mobile user initiates a new class I connection in cell *A*. In the proposed scheme, bandwidth is allocated to the connection in the current cell *A*, and bandwidth is reserved in neighboring cells *B*, 1, 2, 6, 7, and 8. When the user moves from cell *A* to cell *B*, the reserved bandwidth (and additional unused bandwidth, if there is any) in cell *B* is used to accommodate the handoff connection, and the reserved bandwidth in cells 1, 2, and 6 (i.e., cells which are not adjacent to cell *B*) is released. At the same time, bandwidth in cells *A*, 7, 8, 9, 10, and 11 (i.e., new neighboring cells) is reserved.

Various algorithms may be used to determine the amount of bandwidth to reserve for class I connections. A simple approach is to reserve the desired amount of bandwidth of class I connection. However, this approach may result in reserving a huge amount of bandwidth of a large number of class I connections may need to be reserved. As an example, refer to Fig. 6 again. Following this approach, it is necessary to reserve in cell *B* the sum of the desired amount of bandwidth of class I connections in cells *A*, 7, 8, 9, 10, and 11. Since it is very unlikely that all class I connections in cells *A*, 7, 8, 9, 10, and 11 will move to cell *B* at the same time, this approach results in inefficient bandwidth usage. A more efficient approach is to reserve only a fraction of the sum, allowing class I connections to share a common pool of reserved bandwidth. The amount of bandwidth to reserve may be calculated as a function of the requested bandwidth by class I connections (e.g., the average or the largest of all of the requested bandwidths from the neighboring cells) or as a function of the number of class I connections (e.g., the number of class I connections in the neighboring cells).

The proposed scheme is summarized in the following pseudo code shown below:

IF new connection THEN
IF desired amount of bandwidth < unused bandwidth THEN
IF Class I THEN
Accept connection
allocate desired amount of bandwidth
If reservation is not successful in all neighboring cells, the new connection is rejected
IF Class II THEN
Accept connection.
Allocate desired amount of bandwidth.

ELSE / Not enough bandwidth */*
Reject the new connection.

After bandwidth is reserved (for class I connections), network conditions may change. Therefore, the proposed scheme adjusts the size of the reserved bandwidth pool based on the current network conditions. It is assumed that each base station constantly monitors the percentage of reserved bandwidth utilization. The algorithm to control the size of reserved bandwidth pool is summarized in the following pseudo code that is controlled by fuzzy logic rules.

IF Class I
IF minimum required bandwidth<=class I reserved bandwidth+unused bandwidth THEN
accept connection.
Allocate min{desired amount of bandwidth, class I reserved bandwidth+unused bandwidth}
reserve bandwidth in new neighboring cells.
/ If reservation is not successful, drop the connection*/*
Release no longer needed bandwidth.
ELSEIF / not possible to allocate minimum*/*
Drop connection.
ELSEIF Class II
IF unused bandwidth> 0 THEN / as long as there is some available bandwidth*/*

Accept connection

Allocate min{desired amount of bandwidth, unused bandwidth}
ELSE / no bandwidth available */*
Drop connection.

There are some key design options for the proposed scheme. First, in reserving bandwidth for class I connections, if user movement pattern are known, more bandwidth maybe reserved in the cells that users are likely to move to. If such patterns are not known, the same amount of bandwidth may be reserved in each of the neighboring cells. Second, in determining the amount of bandwidth to reserve, various functions can be used. For instance, the amount of bandwidth to reserve may be calculated based either on the requested bandwidth of existing class I connections or on the number of existing class I connections.

5 SIMULATION MODEL DESCRIPTION AND PARAMETER VALUES

The model we used for simulation is composed of 10 cells, each cell keeping contact with six neighboring cells at most. Each cell contains a base station, which is responsible for the connection setup and teardown of new connections and handoff connections, as well as the reservation of bandwidth in neighboring cells. Each cell has a maximum bandwidth capacity of *B* bits/s. A general structure of the proposed model is shown in Fig. 7

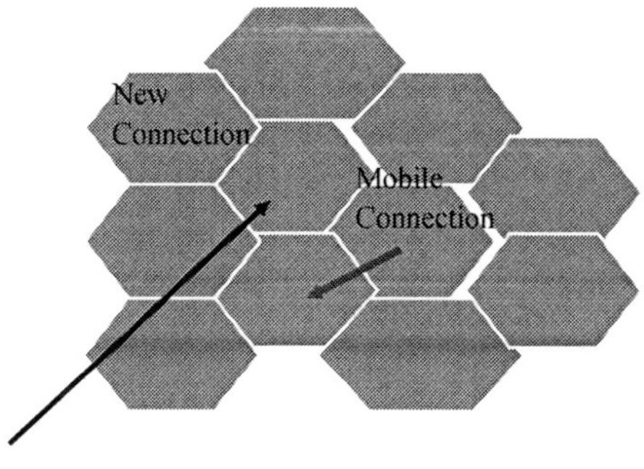

Fig. 7. The cellular system with two types of traffic connections

A zoom out in one of the cells will look like what is shown in Fig. 8

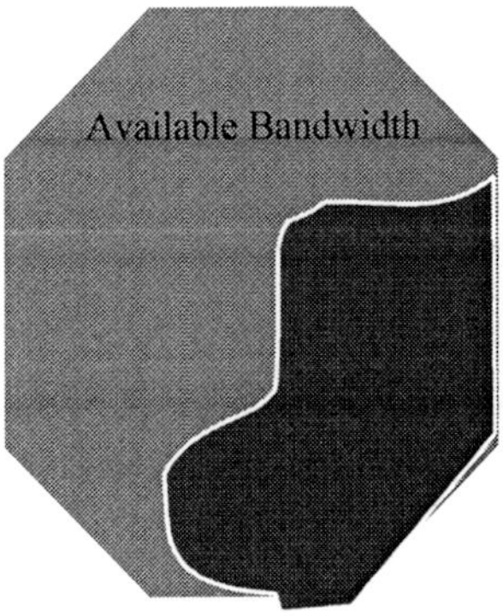

Fig. 8. A zoom out in one of the cellular system cells

To guarantee QoS for class I connections, bandwidth has to be reserved in adjacent cells where the existing connection take place. As we can see from the Fig. 8 above, each cell is divided into two parts; available bandwidth and reserved bandwidth, the total bandwidth of each cell will be the sum of the available and reserved bandwidths. When a new class I connection takes place in one of the cellular system cells, a bandwidth pool of 10 Mb/s is reserved in the adjacent cells.

The choice of 10 Mb/s of a nominal reserved bandwidth has been chosen since all class I connections (used in the simulation) have a Minimum Cell Rate (MCR) of 10 Mb/s. Thus, by allocating the 10 Mb/s reserved bandwidth in adjacent cells, we make sure that if a mobile connection moves to one of the adjacent cells, it would not experience any lack of the QoS parameters that has been agreed upon when the connection was accepted. If we leave the reserved bandwidth in adjacent cells as it is without monitoring its utilization, the cell bandwidth utilization might not be as efficient as it should be. This is because such reserved bandwidth might not be used if none of the existing connections try to move to the neighboring cells.

Such possible inefficient reserved bandwidth utilization will severely affect the connection admission control algorithm performance. The reason for this is that if the reserved bandwidth in the cellular system is not dynamically allocated and monitored, the available bandwidth in each cell in the cellular system will keep decrementing with time. Thus, when CAC would like to accept a new connection or move one of the existing class II connections to another cell, it might get blocked or dropped, respectively, due to the unavailable bandwidth in the desired cell. It is important to mention that mobile class II connections use the available bandwidth in each cell rather than using the reserved bandwidth. The reserved bandwidth in each cell can only be used by class I mobile connections. This point is very important while analyzing the simulation results.

Fuzzy logic rules are used to monitor and adjust the reserved bandwidth in each cell. Thus, in the simulation there is a fuzzy logic controller implemented for each cell in the cellular system in a distributed manner. The fuzzy logic controller will compare the reserved bandwidth utilization with a reference point that can be changed during the simulation. According to the result of the comparison between the reference point and the current reserved bandwidth utilization, some of the rules in the fuzzy controller rule-base will be fired and take action to adjust the reserved bandwidth and thus the available bandwidth.

It is important to mention that the fuzzy controller will not always decrement the reserved bandwidth from the nominal 10 Mb/s that was initially allocated. If the current reserved bandwidth in a cell is not sufficient to the demand of mobile class I connections, the fuzzy controller will increase the reserved bandwidth at the cost if decreasing the available bandwidth to compensate the mobile connections demand.

Two types of new connections are assumed in the simulation: a user connection, which is initiated by a mobile user, and a network connection, which is initiated by a mobile user. The interarrival times of new connection requests are assumed to follow a geometric distribution with mean $1/\lambda_N$ for network connections and mean $1/\lambda_U$ for user connections. It is also assumed that each connection may experience multiple handoffs during its lifetime.

In order to represent various multimedia applications, six different application groups are assumed based on the connection duration, bandwidth requirements, and class of service (class I or class II). The different application groups include constant bit rate (CBR), variable bit rate (VBR), and data traffic sources (unspecified bit rate-UBR). The following table shows six application groups used in the simulation. These are typical applications seen on existing networks. The bandwidth required for VBR and UBR connections is assumed to follow a geometric distribution between the minimum and maximum values shown in Table 1. In the simulation, it is assumed that new connections from all six applications groups are generated with equal probability.

6 SIMULATION RESULTS

The simulation results focused on two main issues. The first simulation analysis is the study of the effect of increasing the reserved bandwidth pool on the dropped and blocked connections for classes I and II. The other simulation analysis issue is the study of the improvement in CAC performance after using fuzzy concepts.

Application Group	Traffic Class	Bandwidth Requirement	Minimum Bandwidth Requirement
1	I	30 Kbps (CBR)	---------
2	I	256 Kbps (CBR)	---------
3	I	1-6 Mbps (VBR)	1 Mbps
4	II	5-20 Kbps (UBR)	5 Kbps
5	II	64-512 Kbps (UBR)	64 Kbps
6	II	1-10 Mbps (UBR)	1 Mbps

Table 1. Samples of the different input traffic connections to the cellular system

6.1 Effect of Size of Bandwidth Reservation Pool

As we can see from Fig. 9, the size of the reserved bandwidth pool in the adjacent cells, to the cell where the connection exists, has a big impact on the number of blocked new connections. When we increase the bandwidth pool size, we are decrementing the available bandwidth in each cell. Thus, the more the allocation of bandwidth for mobile connections is available, the less the probability of new connections to be accepted by CAC.

We observe from Fig. 9 that the cellular system experiences an increase in the total number of blocked new connections when the bandwidth pool is increased. Also, we observe that both class I and II experience the same increase in blocked new connections but with a different effect. As described earlier, class I connections need QoS guarantees in adjacent cells when the existing connection moves to one of them. Thus, CAC job of allocating bandwidth in adjacent cells for class I connections will get harder since less available bandwidth exists when we increase the bandwidth pool in adjacent cells. CAC for class II new connections will not experience the same situation since CAC concern is to allocate bandwidth only in the cell where the new connection would like to take place. This will make the possibility for a blocked class II connection less than the possibility for a class I connection. This explains the reason why the total blocked new connections for class I traffic is more than that for class II traffic for the same bandwidth pool size.

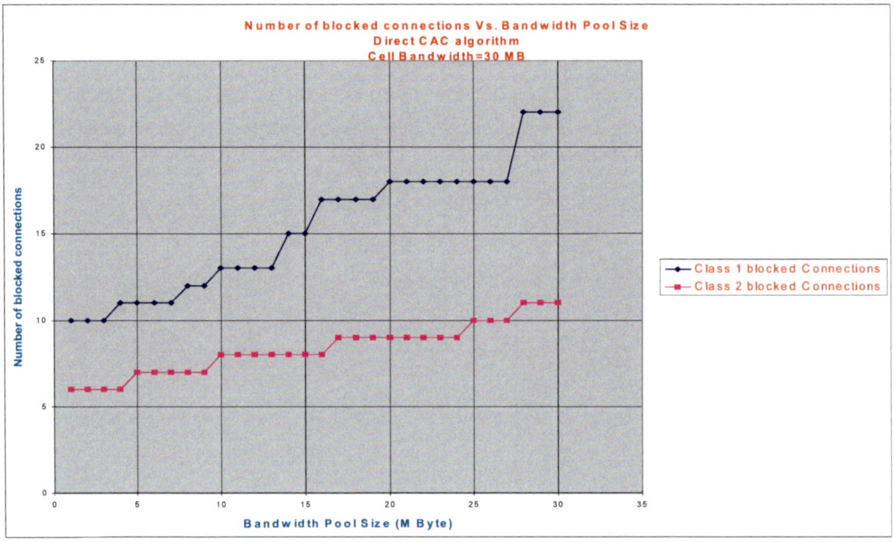

Fig. 9. Number of blocked connection for classes I, II vs. Bandwidth pool size.

Fig. 10 describes the effect of increasing the reserved bandwidth pool size in adjacent cells on the total number of dropped mobile connections. We notice that increasing the bandwidth pool size decrements the total number of dropped connections for class I traffic. When we increase the allocated bandwidth pool in adjacent cells, we are increasing the possibility for CAC to move the mobile connection to the new cell and guaranteeing the required QoS. The possibility for the QoS guarantees increases when the bandwidth pool size for mobile connections increases. It is interesting to notice that class II mobile connections experience a drop in the total number of dropped connections due to bandwidth pool size increase. This is normal since CAC for class II mobile connections use the available bandwidth in the adjacent cells rather the bandwidth pool to decide whether the connection is dropped or moved to a new cell. So, when we increase the bandwidth pool size, we are decreasing the available bandwidth in adjacent cells, and hence increasing the possibility for dropped mobile calls II connections.

Fig. 11 combines both the total blocked (new) and dropped (mobile) connections performance for class I traffic due to bandwidth pool size increase. We can notice from Fig. 11 that the total number of dropped connections starts to decrease when the bandwidth pool size is greater than 10 Mb/s. This is explained due to the reason that all class I connections have a Minimum Cell Rate (MCR) of 10 Mb/s. If the bandwidth pool size is less than MCR, all the mobile connections moving to new cells will be dropped because the allocated reserved bandwidth pool is less than the MCR, and thus does not guarantee the QoS contract that was agreed upon at the moment when the connection was first accepted.

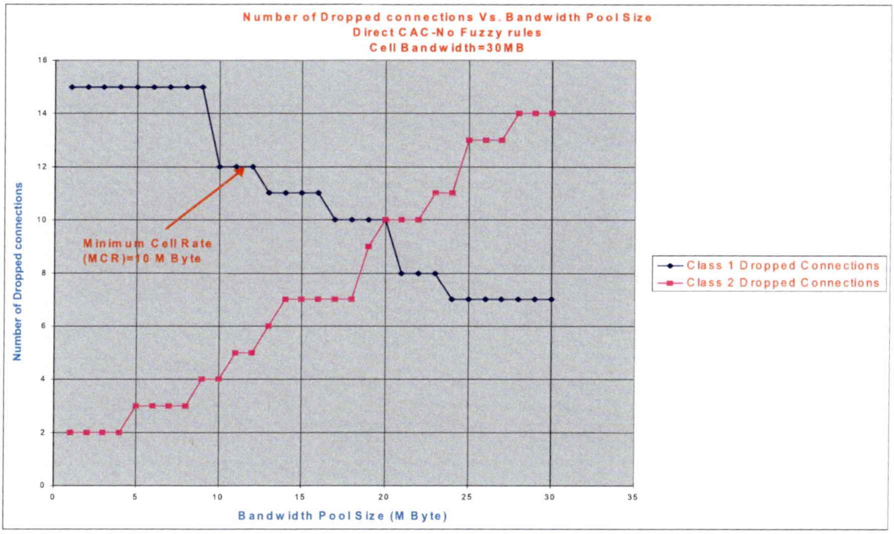

Fig. 10. Dropped mobile connections for class I and II Vs. bandwidth pool size.

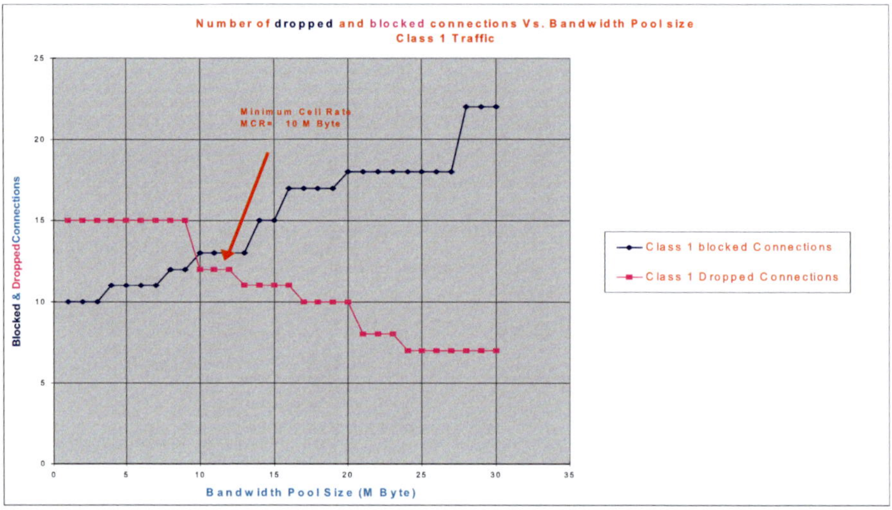

Fig. 11. Dropped and blocked connections for class I traffic Vs. Bandwidth pool size.

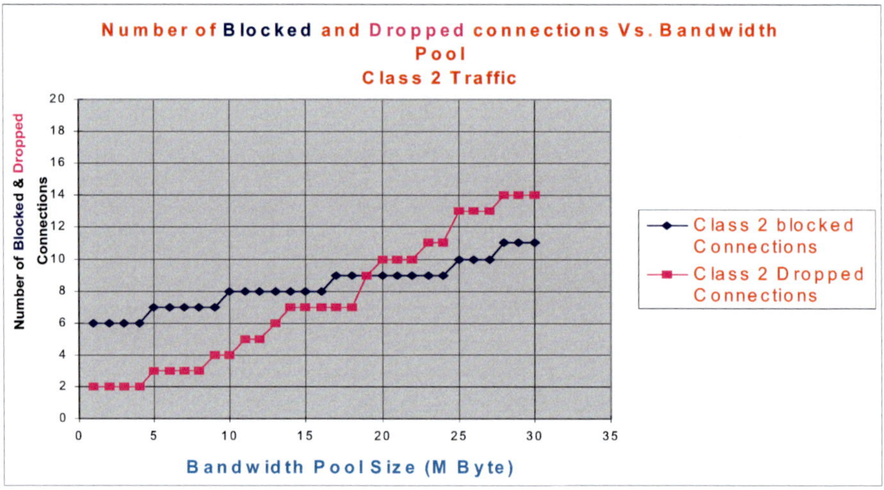

Fig. 12. Dropped and blocked connections for class II traffic Vs. Bandwidth pool size.

Fig. 12 combines both the effect of pool size increase on the dropped and blocked connections for class II traffic. As seen from the Figure, both the total blocked and dropped connections of class II experience an increase when the reserved bandwidth pool increases. The reason behind the increase in the blocked connections is due to the decrease in available bandwidth of adjacent cells as a result of the increase of bandwidth pool. As we mentioned earlier, CAC algorithm for new class II connections will try to allocate them in the available bandwidth not the reserved bandwidth since the later is for mobile class I connections.

Also, the increase in the dropped connection despite the increase in the bandwidth reserved pool can be explained using the same reason above. The bandwidth pool is reserved for mobile class I connections and not for class II mobile connections. Thus, it is obvious to experience an increase in the dropped class II connections due to the decrease in the available bandwidth.

6.2 Effect of Learning Fuzzy CAC on Blocked and Dropped Connections

As was described earlier, both direct and learning fuzzy logic were combined with the CAC algorithm to study the effect of applying fuzzy rules into the performance of CAC and thus the number of blocked and dropped connections for both classes. The problem with CAC alone without any intelligence is the fixed allocation of bandwidth pool in adjacent cells. This fixed bandwidth allocation is not recommended in a very dynamic and changing system conditions like a cellular system.

If the utilization of reserved bandwidth pool in adjacent cells is not monitored and adjusted according to network conditions. The cells bandwidth will experience a great decrease in bandwidth utilization and thus affecting the performance of CAC. Of course when CAC does not perform as required, unaccepted total number of dropped and blocked connections will be experienced. Thus, the main reason behind using fuzzy logic is to enhance and smoothen the performance of CAC for a better network bandwidth utilization and thus more accepted new and mobile connections.

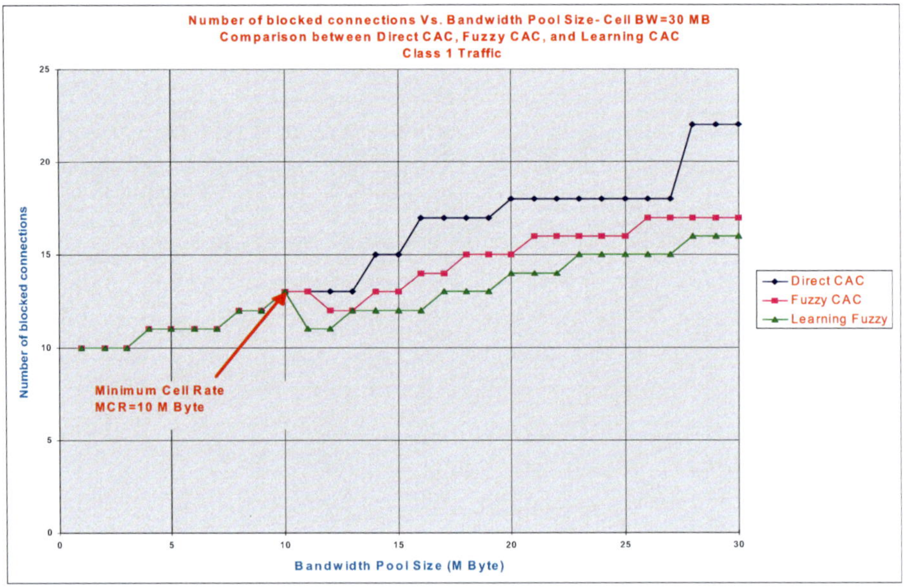

Fig. 13. Total number of blocked class I connections for direct, fuzzy, and learning fuzzy CAC.

Fig. 13 describes the effect of fuzzy logic on CAC performance over class I blocked connections. Notice from this Figure, that the direct CAC, fuzzy CAC, and the learning fuzzy CAC have the same performance for bandwidth pool size less than 10 Mb/s, i.e., they have the same total number of blocked new connections. This is normal because 10Mb/s is the Minimum Cell Rate (MCR) for class I connections. Since any bandwidth pool size less than MCR would not guarantee any type of class I traffic sources, neither fuzzy CAC nor learning fuzzy CAC would have any effect on the total number of blocked connections.

As the reserved pool size is increased above MCR, we notice a significant CAC improvement in performance over the total number of blocked connections. For any reserved pool size greater than MCR, fuzzy CAC produces less total number of blocked connections than direct CAC. This is explained due to the effect of fuzzy rules on the nominal reserved bandwidth pool.

In any cell, after the 10 Mb/s nominal bandwidth is reserved, fuzzy rules of the fuzzy CAC will keep monitoring the bandwidth pool utilization. If the allocated bandwidth is not utilized or used by any of the moving class I connections, fuzzy rules will decrement the nominal reserved bandwidth and hence increase the available bandwidth. This of course will cause CAC to accept bigger number of new connections to the cellular system and hence the total number of blocked connections will be less.

It is worth to mention that fuzzy rules role is not only to decrement the nominal bandwidth pool, but also to increase the bandwidth pool size if the current bandwidth pool size is not sufficient or enough to guarantee the QoS of mobile class I connections.

We can also notice from Fig. 13 that the learning fuzzy CAC has a better (less) number of total blocked connections than both direct CAC and fuzzy CAC for pool size greater than MCR. As was mentioned earlier about the advantage of the learning fuzzy logic rules, it has a faster response to the dynamic network conditions. Thus, the leaning fuzzy CAC will sense any dynamic changes in the cell reserved bandwidth utilization faster than both fuzzy CAC and direct CAC. As a result of this, learning fuzzy CAC will be able to take action and decrement or increment the reserved bandwidth pool size faster than fuzzy CAC. This leads to a better cell bandwidth utilization and hence less total number of blocked connections.

Fig. 14 illustrates the enhancement of fuzzy rules on CAC performance over the total dropped class I connections. As was described earlier, both fuzzy CAC and learning fuzzy CAC have significant improvements over direct fuzzy when the reserved pool size is increased above MCR. We can notice from Fig. 14 that both fuzzy CAC and learning fuzzy CAC performance causes less total number of dropped class I connections for any pool since greater than MCR. This can be explained due to the efficient utilization of the reserved bandwidth pool.

6.3 Effect of Learning Fuzzy CAC on Reserved Bandwidth Utilization

One of the great advantages of using fuzzy logic or the leaning fuzzy logic rules with CAC is the online and dynamic monitoring and adjustment of reserved bandwidth allocation. Fig. 15 is a clear illustration of such advantage. This Figure shows that direct CAC has a fixed reserved bandwidth allocated at all neighboring cells at the time of new connection acceptance. If this fixed allocated bandwidth is not used by any mobile connection in the neighboring cells, the reserved bandwidth utilization will be low and hence will affect CAC performance on the total new and mobile connection acceptance.

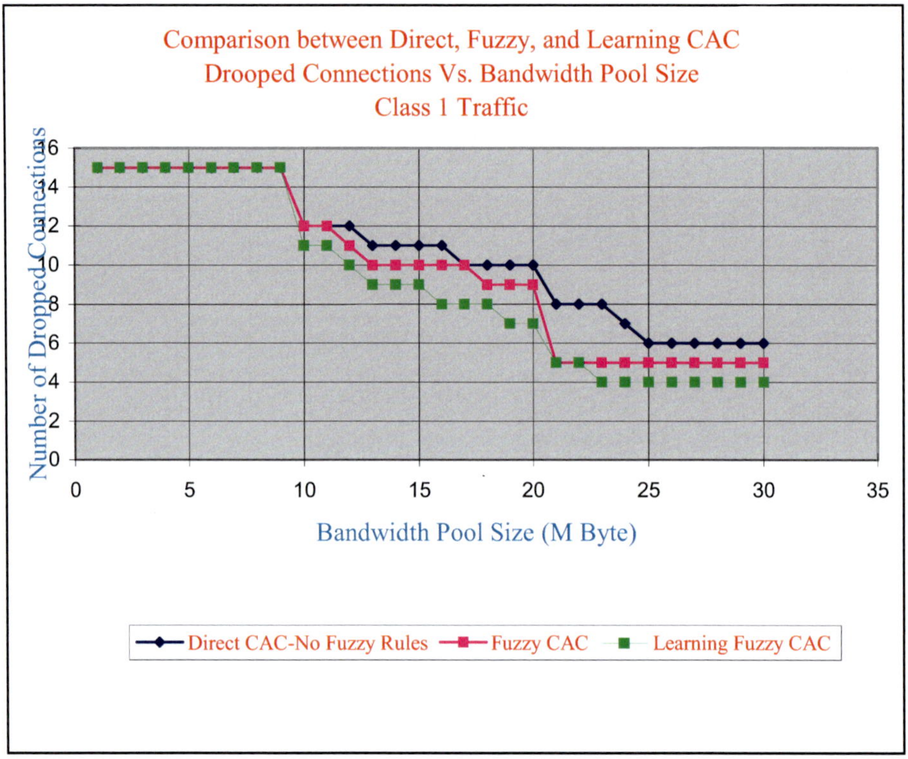

Fig. 14. Total number of blocked class I connections for direct, fuzzy, and learning fuzzy CAC.

Note from Fig. 15 that learning fuzzy CAC has better reserved bandwidth adjustment and monitoring procedure than both direct CAC and fuzzy CAC. After the fixed reserved bandwidth has been allocated by CAC, fuzzy logic rules will monitor the reserved bandwidth utilization. If the reserved bandwidth utilization is low, fuzzy logic rules will decrease the reserved bandwidth and hence increase the available bandwidth in each neighboring cell. Also, fuzzy logic rules will increase the current reserved bandwidth if its not sufficient for a new mobile class I connection, i.e., less than MCR.

Learning fuzzy logic has a better monitoring and adjustment performance than direct fuzzy rules. As we notice from Fig. 15 that learning fuzzy rules sense the dynamic changes of the network conditions (cell bandwidth utilization) faster than direct fuzzy rules. This provides a better bandwidth utilization and hence improves the total number of accepted new and mobile connections. Meaning that learning fuzzy rules will increase or decrease the increment or decrement in reserved bandwidth according to the current changes in reserved bandwidth utilization.

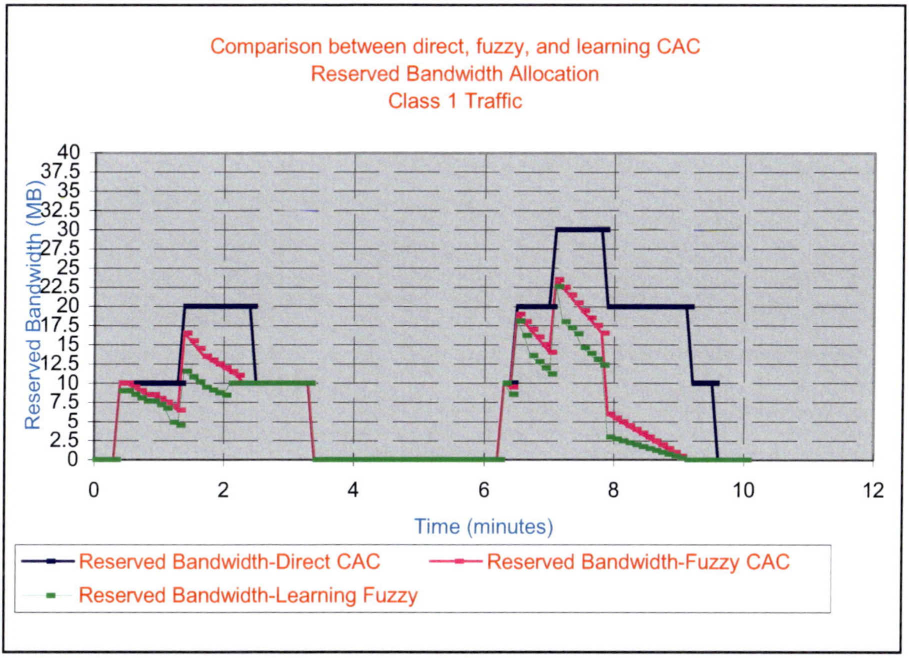

Fig. 15. Comparison between direct, fuzzy, and learning fuzzy CAC for class I traffic reserved bandwidth allocation.

7 Conclusion

Using fuzzy logic rules improves the reserved bandwidth utilization and hence decreases the blocked and dropped connections. Through continuous monitoring and adjustment of the reserved bandwidth pool size in adjacent cells, fuzzy logic rules were able to allocate the appropriate bandwidth for the current network conditions. We have seen that through this continuous bandwidth adjustment, CAC performance produced a less total number of blocked and dropped connections for both traffic classes.

Learning fuzzy logic produced a better CAC performance. We have noticed from the simulation results that learning fuzzy logic rules sense the network dynamic changes faster. This provides faster decision-making time for adjusting the reserved bandwidth to cope with the new conditions that will result is less total dropped and blocked connections of both traffic types.

In this paper, we proposed a learning fuzzy connection admission control fro two classes of ATM traffic. It was noticed from the simulation that increasing the bandwidth pool size decreases the allowed number of new connections, i.e., increases the number of blocked connections. So, when deciding the size of the reserved bandwidth pool, we should determine what is our main concern for CAC. If the main concern is to decrease the total number of dropped connections, then bandwidth pool size should be greater than MCR. On the other hand, if the main concern is to decrease the total blocked new connections, then we have to fix the reserved allocated bandwidth to MCR.

An interesting observation was noticed regarding the effect of pool size on class II dropped connections. It was observed that increasing the reserved bandwidth pool size increases the dropped mobile class II connections. This is normal since class II mobile connections do not use the reserved bandwidth in neighboring cells, but they use the available bandwidth. Since increasing the reserved bandwidth decreases the available bandwidth in neighboring cells, the probability for mobile class II connections acceptance in neighboring cells will be less, i.e., larger total number of dropped class II connections.

REFERENCES

[1] Eckberg, *B-ISDN/ATM Traffic and Congestion Control*, *IEEE Network Magazine*, Vol. 13 (1992), no. 9, pp. 9-14.

[2] M. Guizani and A. Rayes, *Designing ATM Switching Networks*, McGraw-Hill, 1999.

[3] H. Oshaki *et al.*, *Rate-based Congestion Control for ATM Networks*, *Computer Communication Review*, April 1995, pp. 56-65.

[4] A. Tanenbaum, *Computer Networks*, 3rd Ed., Prentice Hall, 1996.

[5] W. Stallings, *High-Speed Networks: TCP/IP and ATM Design Principles*, Prentice Hall, 1998.

[6] Layne, J. and K. Passino, *Fuzzy Model Reference Learning Control*, Proceedings of the 1st IEEE Conference on Control Applications, 1992, Dayton, Ohio, USA.

[7] K. Passino and S. Yurkovich, *Fuzzy Control: Theory and Applications*, Addison Wesley, 1997.

INDEX

R

S

T

U

V

W